HIGH TEMPERATURE SUPERCONDUCTIVITY FROM RUSSIA

SERIES ON PROGRESS IN HIGH TEMPERATURE SUPERCONDUCTIVITY

Published

Vol. 1 — Proceedings of the Adriatico Research Conference on High Temperature Superconductors
(*eds. S. Lundqvist, E. Tosatti, M. Tosi and Yu Lu*)

Vol. 2 — Proceedings of the Beijing International Workshop on High Temperature Superconductivity
(*eds. Z. Z. Gan, G. J. Cui, G. Z. Yang and Q. S. Yang*)

Vol. 3 — Proceedings of the Drexel International Conference on High Temperature Superconductivity
(*eds. S. Bose and S. Tyagi*)

Vol. 4 — Proceedings of the 2nd Soviet-Italian Symposium on Weak Superconductivity
(*eds. A. Barone and A. Larkin*)

Vol. 5 — Proceedings of the IX Winter Meeting on Low Temperature Physics — High Temperature Superconductors
(*eds. J. L. Heiras, R. A. Barrio, T. Akachi and J. Tagüeña*)

Vol. 7 — Chemical and Structural Aspects of High Temperature Superconductors
(*ed. C. N. R. Rao*)

Vol. 8 — World Congress on Superconductivity
(*eds. C. G. Burnham and R. Kane*)

Vol. 9 — First Latin-American Conference on High Temperature Superconductivity
(*eds. R. Nicolsky, R. A. Barrio, R. Escudero and O. F. de Lima*)

Vol. 12 — High Temperature Superconductivity and Other Related Topics — 1st Asia-Pacific Conference on Condensed Matter Physics
(*eds. C. K. Chew et al.*)

Vol. 14 — Proceedings of the Adriatico Research Conference and Workshop on Towards the Theoretical Understanding of High Temperature Superconductors
(*eds. S. Lundqvist, E. Tosatti, M. Tosi and Yu Lu*)

Vol. 15 — International Symposium on New Developments in Applied Superconductivity
(*ed. Y. Murakami*)

Vol. 16 — Proceedings of the Srinagar Workshop on High Temperature Superconductivity
(*eds. C. N. R. Rao et al.*)

Vol. 17 — Proceedings of High-T_c Superconductors: Magnetic Interactions
(*eds. L. H. Bennett, Y. Flom and G. C. Vezzoli*)

Vol. 18 — Proceedings of the Tokai University International Symposium on the Science of Superconductivity and New Materials
(*ed. S. Nakajima*)

Forthcoming

Vol. 10 — Macroscopic Theories of Superfluidity and Superconductivity
(*ed. A. A. Sobyanin*)

Vol. 13 — Applications of High Temperature Superconductors
(*ed. C. Y. Huang*)

Vol. 19 — Superconductivity and Applications — Taiwan International Symposium on Superconductivity
(*eds. H. C. Ku and P. T. Wu*)

Vol. 20 — Proceedings of the X Winter Meeting on Low Temperature Physics — High Temperature Superconductors
(*eds. T. Akachi, J. A. Cogordán and A. A. Valladares*)

Vol. 21 — Proceedings of the International Seminar on High Temperature Superconductivity
(*eds. N. N. Bogolubov, V. L. Aksenov and N. M. Plakida*)

Progress in High Temperature Superconductivity — Vol. 11

HIGH TEMPERATURE SUPERCONDUCTIVITY FROM RUSSIA

Editors

A.I. Larkin
N.V. Zavaritsky

World Scientific
Singapore • New Jersey • London • Hong Kong

Published by

World Scientific Publishing Co. Pte. Ltd.,
P O Box 128, Farrer Road, Singapore 9128
USA office: 687 Hartwell Street, Teaneck, NJ 07666
UK office: 73 Lynton Mead, Totteridge, London N20 8DH

HIGH TEMPERATURE SUPERCONDUCTIVITY FROM RUSSIA

Copyright @ 1989 by World Scientific Publishing Co. Pte. Ltd.

All rights reserved. This book, or parts thereof, may not be reproduced in any form or by any means, electronic or mechanical, including photocopying, recording or any information storage and retrieval system now known or to be invented, without written permission from the Publisher.

ISBN 9971-50-798-6

Printed in Singapore by Utopia Press.

FOREWORD

This volume covers up-to-date ideas associated with the studies of high-T_c superconductivity. The book largely comprises contributions presented at the 25th All-Union Conference on Low Temperature Physics (Leningrad, 25-27 October 1988). The editors have chosen about one-sixth of the presented High-T_c articles.

At present there is no unified point of view on the nature of high-T_c superconductivity and our book presents diverse points of view on this problem. Besides, a number of works is devoted to a theoretical analysis of concrete properties of high-T_c superconductivity.

Experimental works discuss the results obtained in the studies of superconducting Bi- and Tl-based compounds as well as the results of thorough investigation of different properties of 123-compounds on the basis of Y-Ba-Cu with different oxygen content. A few articles are devoted to SQUIDS functioning at nitrogen temperatures and to their applications in physical researches.

Anatoly Larkin and Nikolai V. Zavaritsky

CONTENTS

Foreword   v

Electronic Structure and "Chemical" Aspects of High-Temperature Superconductivity   1
   D.I. Khomskii

The Holes on Oxygen Sites and BCS Physical Picture in High-temperature Superconductors   22
   G.M. Eliashberg

Polaron Theory of High-T_c Superconductors   54
   A.S. Alexandrov

Mott Transition: Low Energy Excitations and Superconductivity   62
   L.B. Ioffe and A.I. Larkin

Electron-exciton Interaction and Local Pairing in High-T_c Superconductors   81
   I.O. Kulik and A.G. Pedan

Time Dependence of Magnetization of High-Temperature Superconductors   93
   V.B. Geshkenbein and A.I. Larkin

The Bernoulli Effect in High-T_c Superconductors   99
   A.N. Omel'yanchuk

Preparation and Study of Tl-based Superconductors   105
   A.I. Akimov, L.Z. Avdeev, S.V. Bogachev, B.B. Boiko,
   M.M. Gaidukov, V.I. Gatalskaya, S.M. Demyanov,
   A.L. Karpei, A.M. Klushin, L.A. Kurochkin,
   Yu.N. Leonovich, L.P. Poluchankina, O.V. Snigirev,
   E.K. Stribuk, and I.M. Starchenko

Superconductivity in the Bi-Sr-Ca-Cu-O and Tl-Ba-Ca-Cu-O
Systems   115

    M.P. Petrov, A.T. Burkov, T.B. Zhukova, A.V. Ivanov,
    N.F. Kartenko, M.V. Krasinkova, A.A. Nechitailov,
    I.V. Pleshakov, V.V. Prokofiev, C.V. Razumov,
    A.I. Grachev, V.V. Poborchy, S.S. Ruvimov,
    S.I. Shagin, V.V. Potapov, A.V. Golubkov, and
    A.V. Prochorov

On Possible Methods of Synthesis and Properties of
Superconductive Thallium Cuprates   132

    V.I. Ozhogin, L.D. Shustov, A.B. Myasoedov,
    N.S. Tolmacheva, A.V. Inyushkin, A.N. Taldenkov,
    N.A. Babushkina, E.P. Krasnoperov, and
    Yu.A. Teterin

Structure Genesis and the Physical Properties of Perovskite-
like Phases   152

    N.E. Alekseevskii, D. Wlosewicz, A.V. Mitin,
    V.I. Nizhankovskii, V.N. Narozhnyi, E.P. Khlybov,
    G.M. Kuzmicheva, I.A. Garifullin, N.N. Garifianov,
    B.I. Kochelaev, A.V. Gusev, G.G. Deviatykh, and
    A.V. Kabanov

Vortex Structure of High-T_c Single Crystals   171

    L.Ya. Vinnikov, I.V. Grigorieva, L.A. Gurevich,
    and Yu.A. Ossipyan

Critical State and Field H_{c1} in $YBa_2Cu_3O_{7-x}$ Single Crystals   181

    M.V. Kartsovnik, V.M. Krasnov, V.A. Larkin,
    V.V. Ryazanov, and I.F. Schegolev

Copper NMR and NQR Studies of the $YBa_2Cu_3O_{7-x}$ High-T_c
Superconductor   187

    E. Lippmaa, E. Joon, I. Heinmaa, A. Vainrub,
    V. Miidel, A. Miller, I.F. Schegolev, I. Furo,
    and L. Mihaly

Scanning Tunneling Spectroscopy of a High-T_c Superconducting
Monocrystal $YBa_2Cu_3O_7$ 201

    A.P. Volodin, M.S. Khaikin, and G.A. Stepanyan

Reflectivity and Raman Spectra of Superconducting and
Non-superconducting $YBa_2Cu_3O_{7-\delta}$ Crystals ... 211
    M.P. Petrov, A.I. Grachev, M.V. Krasinkova,
    N.F. Kartenko, A.A. Nechitailov, V.V. Poborchii,
    V.V. Prokofiev, S.S. Ruvimov, and S.I. Shagin

IR Reflectivity Spectra of $Y_1Ba_2Cu_3O_{7-x}$ Single Crystals ... 224
    A.V. Bazhenov, A.V. Gorbunov, and V.B. Timofeev

Effect of Oxygen Concentration on Transport Properties of
$Y_1Ba_2Cu_3O_x$, ($6 < x < 7$) ... 238
    A.V. Samoilov, A.A. Yurgens, and N.V. Zavaritsky

The Effect of a High Pressure up to 210 kbar upon the
Superconductivity of Monocrystalline and Ceramic Specimens
of $Y_1Ba_2Cu_3O_{7-x}$ ... 250
    I.V. Berman, N.B. Brandt, I.L. Romashkina,
    V.I. Sidorov, and C.Y. Han

Acoustic Characteristics and Peculiarities of the Lattice
Vibration Spectrum in $La_{2-x}Sr_xCuO_4$ ($x = 0; 0,2$) and
$YBa_2Cu_3O_y$ ($y = 6; 7$) ... 263
    V.S. Klochko, V.I. Makarov, V.F. Tkachenko,
    A.P. Voronov, and N.V. Zavaritsky

Effects of Localisation in Atomic-disordered High-T_c
Superconductors ... 270
    B.N. Goshchitskii, S.A. Davydov, A.E. Karkin,
    A.V. Mirmelstein, M.V. Sadovskii, and V.I. Voronin

High-T_c Ceramic Weak Links, rf-SQUIDs and their Applications ... 281
    N.V. Zavaritsky and V.N. Zavaritsky

High-T_c Superconducting Thin Film Weak Links: Josephson
Effect and Macroscopic Quantum Interference at 4 K and 77 K ... 300
    A.I. Golovashkin, A.L. Gudkov, S.I. Krasnosvobodsev,
    L.S. Kuzmin, K.K. Likharev, Yu.V. Maslennikov,
    Yu.A. Pashkin, E.V. Pechen, and O.V. Snigirev

ELECTRONIC STRUCTURE AND "CHEMICAL" ASPECTS OF HIGH-TEMPERATURE SUPERCONDUCTIVITY

D.I.Khomskii

P.N.Lebedev Physical Institute of the Academy of Sciences,
Moscow, USSR

Contents

1. Introduction
2. Copper ions: electronic structure
3. Copper ions: interactions, valence
4. Oxygen ions: energy level structure
5. p-d hybridization, bands and exchange interaction
6. "Neutral oxygen" and superconductivity

1. Introduction

When discussing high-temperature superconductivity (HTSC) the first question one encounters is what is so specific in copper oxides which makes them high-temperature superconductors? There were many suggestions concerning possible nature of HTSC. It is plausible in principle that the mechanism of superconductivity in these systems is of a rather traditional type. However it may turn out that it is radically different from the one known before. But in any case one should answer the question why just in these compounds is HTSC realized, what is necessary for new mechanisms to be effective or why the usual mechanisms give here T_c 5 times higher than before? To this end we have to discuss in more details the specific features of copper metallooxides especially their electronic structure.

In this report I will discuss certain aspects of electronic structure and properties of the known HTSC's based on copper although some topics covered may be relevant also to new "not so high" HTSC's like $Ba_{1-x}K_xBiO_3$. I shall mostly dwell upon those properties which are connected with the behaviour of constituent atoms and ions; it is in this sense that I use the term "chemical aspects". Correspondingly I shall use "chemical" language and shall operate mostly with the picture of localized "atomic" states not only for copper d-states for which this picture is probably adequate but also for oxygen p-states. Whenever it is necessary (and when possible) the connection will be established with the band picture.

We shall give our discussion using mostly La_2CuO_4 as a typical example. It seems now that the situation in other Cu-containing HTSC's is in essence the same.

2. Copper ions: electronic structure

The main structural element in all the known HTSC's is CuO_2 plane. Cu ions are located in a tetragonal surrounding: elongated O_6 octahedra in "La" compounds or 5-fold coordination in "Y" or "Bi" systems. The splitting of one-electron d-levels in this case is shown in fig.1. The typical values of the splitting for CuO_6 clusters are: $\Delta E \simeq 1$ eV; $\Delta \varepsilon = \varepsilon_{x^2-y^2} - \varepsilon_{z^2} \simeq 0.7$ eV /1/. Note that for square coordination (copper in chains in "123") tetragonal splitting of d-levels may turn out so strong that the order of levels will be reversed: d_{z^2} and d_{xy} - levels may cross and uppermost levels will be $d_{x^2-y^2}$ and d_{xy} /2/ (fig.1d).

Formal valence state of copper in La_2CuO_4 is Cu^{2+}; the same valence is now believed to exist also in doped La system and to a large extent in Y, Bi and Tl HTSC's. Cu^{2+} is a well-known Jahn-Teller ion. The splitting of e_g-levels $d_{x^2-y^2}$ and d_{z^2} is here due both to the underlying lattice structure and to the Jahn-Teller effect. Thus in La_2NiO_4 which contains non-Jahn-Teller ions Ni^{2+} the oxygen octahedra are already elongated along C-axes apparently due to the La ions but in La_2CuO_4 with strong Jahn-Teller ion Cu^{2+} this local distortion of O_6-

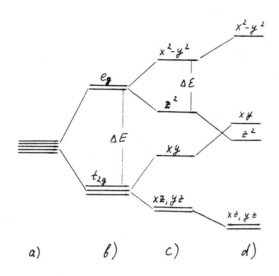

Fig.1. The scheme of crystal field splitting of d-levels. a) Isolated ion; b) Cubic surrounding (regular octahedron); c) Tetragonal surrounding (elongated octahedron); d) 4-fold (square) coordination.

octahedron is twice as large due to Jahn-Teller effect /1/. Moreover it seems that the very existence of rather strange crystal structures of "123" or "Bi" types is determined by the Jahn-Teller character of Cu^{2+} ions. It is well known that just due to this factor Cu^{2+} does often exist in five-fold (pyramid) or four-fold (square) coordination /2/. As the energy of this Jahn-Teller stabilization is of the order of ~1 eV it may well provide a mechanism of stabilization of these curious structures (see also /4/).

On the other side say 3-fold or 2-fold coordination is very untypical for Cu^{2+}, but Cu^{1+} ions do often have two neighbouring ligands. Presumably these tendencies are responsible for the type of the ordering of oxygen vacancies in $YBa_2Cu_3O_{7-y}$. If disordered, oxygen vacancies in chains would produce wrong coordination for certain Cu ions (3, or 4 but not forming a square etc.). The same would be true also if these vacancies would order in each chain e.g. occupying every 3-d site (for $y = 1/3$); such type of ordering was sometimes suggested at the initial stage of investigations of these compounds. But it is established now that equilibrium state corresponds to ordering of "full" and "empty" chains (fig.2). It is clear that just within this structure the valence and coordination rules mentioned above are fulfilled.

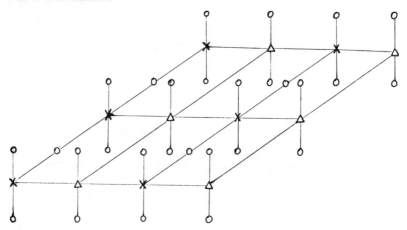

Fig.2. Ordering of oxygen vacancies in chains in $YBa_2Cu_3O_{7-y}$ ($y = 0.5$); $x - Cu^{2+}$, $\triangle - Cu^{1+}$, o - O-ions.

In Cu^{2+} ion there is one hole in d-shell which occupies $d_{x^2-y^2}$ orbital. Quite often when discussing the properties of these systems one retains only this orbital and ignores remaining d-states. Thus it was believed initially that the extra hole introduced say by doping La_2CuO_4 by Sr goes into this orbital transforming $Cu^{2+}(t_{2g}^6 d_{z^2}^2 d_{x^2-y^2}^1)$ into $Cu^{3+}(t_{2g}^6 d_{z^2}^2)$. Standard one-electron approach e.g. band-structure calculations always give such a result. If we ignore for a moment the effects connected with the strong Coulomb d-d interaction (Hubbard terms) which apparently suppress the formation of Cu^{3+} state (see below) and suppose that such a state may be really formed, then even in this case one-electron picture may break down. The extra hole may occupy not the highest $d_{x^2-y^2}$ level but the next d_{z^2}-level instead. We lose in this case an energy $\Delta\varepsilon$ (~0.7 eV), see fig.1, but we gain an energy due to intraatomic Hund rule coupling. Indeed Cu^{3+} state with the configuration $t_{2g}^6 d_{z^2}^2$ would have spin S=0; but if extra hole goes into z^2-orbital then the configuration $t_{2g}^6 d_{z^2}^1 d_{x^2-y^2}^1$ would correspond to S=1 state, and as an intraatomic (Hund) exchange stabilizing such a state is typically J_H ~1 eV this gain may overcome one-electron loss.

We note also that the existence of other d-levels besides $d_{x^2-y^2}$ may prove important for certain properties of these compounds. Thus the mixing of $d_{x^2-y^2}$ and $d_{xz,yz}$-levels may naturally explain /3/ the tetragonal-orthorombic structural phase transition observed in La_2CuO_4. This transition consists mainly in antiphase rotations of CuO_6 octahedra along [110] axis which can be described by the distortions of $E_{xz,yz}$-type. These distortions have no diagonal matrix elements within (x^2-y^2)-manifold, $\langle x^2-y^2 | \hat{E}_{xz,yz} | x^2-y^2 \rangle = 0$, which means that electron-phonon interaction of (x^2-y^2)-electrons alone can not provide a mechanism for such a transition. But the nondiagonal matrix elements are nonzero:

$$\langle x^2-y^2 | E_{xz,yz} | xz,yz \rangle \neq 0 \qquad (1)$$

The mixing of these orbitals and splitting of corresponding energy levels decrease electronic energy and can in principle bring about the observed structural phase transition which may be viewed as a pseudo-Jahn-Teller one. The critical temperature of this transition is given by the expression /3/

$$T_{str} \sim \tilde{\varepsilon} / \text{arc th}(\tilde{\varepsilon}/\frac{g^2}{\omega}) \qquad (2)$$

where $\tilde{\varepsilon} = \varepsilon_{x^2-y^2} - \varepsilon_{xz,yz}$, ω is the frequency of $E_{xz,yz}$ -

vibrations and g is the electron-phonon coupling constant. One can see that $T_{str} \neq 0$ for $g^2/\omega > \tilde{\varepsilon}$ i.e. the coupling with $E_{xz,yz}$-mode should be strong enough.

It is clear also that the same mixing with $d_{xz,yz}$-orbitals would occur for d_{z^2}-states. If, as it is sometimes suggested, due to stronger Coulomb interaction (less screening) d_{z^2}-levels turn out to be close enough to $d_{x^2-y^2}$ then their thermal occupation would increase the distortion. As temperature drops this contribution would be switched off; this may explain nonmonotonous dependence of orthorombic distortion with temperature /21/ and even possible transition back to tetragonal phase for low T (F.Steglich, private communication).

It is evident also that this mechanism of structural transition should operate e.g. for La_2NiO_4 (Ni^{2+}, $t_{2g}^6 d_{x^2-y^2}^1 d_{z^2}^1$), but not for La_2MnO_4 (all d-levels of Mn^{2+} are singly occupied). The corresponding phase transition in La_2NiO_4 was indeed observed /19/; it should be interesting to check whether such a transition would occur in La_2MnO_4 (if this compound does exist).

There exist in principle other important consequences of the rich structure of d-levels. Thus the interaction of conduction electrons with intraatomic d-d transitions was invoked /4,20/ as a possible source of superconducting pairing (the holes in this model are believed to be moving in oxygen p-band, see below). The numerical estimates made by Weber /4/ show that this mechanism can in principle give high enough values of T_c. It is not clear however why this mechanism does not operate say in La_2NiO_4. The arguments presented in /4/ that in Ni^{2+} ions both $d_{x^2-y^2}$ and d_{z^2} levels are singly occupied and d-d transition between them is ineffective are inconclusive: the transitions between these levels and $d_{xy,xz,yz}$ could play the same role leading to an effective electron-electron attraction both in doped La_2CuO_4 and La_2NiO_4.

3. Copper ions: interactions, valence

Up to now we discussed mostly one-electron aspects and ignored electron-electron interaction (exept Hund coupling).

In this scheme pure La_2CuO_4 should be a metal (it contains odd number of electrons per cell). However it is established now that stoichiometric La_2CuO_4 is an insulator with large enough energy gap~2-3 eV [4-6]. Apparently it is connected with the Mott-Hubbard localization of d-electrons due to a strong Coulomb d-d repulsion. In the simplest model with nondegenerate $d_{x^2-y^2}$-level one could use standard Hubbard model

$$H = t \sum_{\langle ij \rangle, \sigma} d^+_{i\sigma} d_{j\sigma} + U_d \sum_i d^+_{i\uparrow} d_{i\uparrow} d^+_{i\downarrow} d_{i\downarrow} \qquad (3)$$

where $d_{i\sigma}$ describe $d_{x^2-y^2}$-holes. The interaction constant U_d is found experimentally [6,7] to be $U_d \sim 7$ eV and the transfer integral t for the transition between neighbouring Cu sites is $t \sim 0.3$ eV *) so that the effective bandwidth of d-band $W = 2zt \sim$ ~2eV < U_d. Therefore pure La_2CuO_4 should be treated as a Mott insulator similar to other typical transition metal oxides like NiO. The existence of antiferromagnetism in La_2CuO_4 with rather high value of $T_N \sim 250$ K and large moment $\mu_{eff} \sim 0.6 \mu_B$ [8] also agree with this picture.

Superconductivity (and metallic conductivity) arises when we substitute La^{3+} by ions like Sr^{2+}. In this way we introduce extra holes into energy bands which apparently are responsible for charge transport and for superconductivity. The situation in $YBa_2Cu_3O_{7-y}$ is the same: for $y \gtrsim 0.6$ the substance is an insulator but it becomes a metal and superconductor for higher oxygen concentration. Thus it seems that formally one could write down the valence state of these systems as $La^{3+}_{2-x}Sr^{2+}_x Cu^{2+}_{1-x} Cu^{3+}_x O^{2-}_4$ or $Y^{3+}Ba^{2+}_2 Cu^{2+}_{2+2y} Cu^{3+}_{1-2y} O^{2-}_{7-y}$ (for $y \leq 0.5$). Sometimes one tries to describe this situation by the pure Hubbard model (3). In this case one should treat the motion of a hole on the antiferromagnetic background. Due to strong interaction of charge and spin degrees of freedom the kinetic energy of a hole is strongly renormalized which can even lead to a sort of "confinement" [22] or to a boson nature of a hole

---
*) Actually d-d transfer occurs via oxygen p-states, see below.

("holons" /23/). This approach is now widely used to discuss different "magnetic" mechanisms of HTSC.

However when we turn to the real systems we see that the situation can hardly be described by the simple Hubbard model. One of the minor complications occurs in 123 compounds. As we mentioned above there are now reasons to believe that when we decrease oxygen content not only some of the formal Cu^{3+} ions are transformed to Cu^{2+} but also Cu^{1+} states appear in chains. Thus in a completely ordered state of $YBa_2Cu_3O_{7-y}$ every y-th chain should be without oxygen i.e. should contain Cu^{1+}. In this case the formal valence state would be not $YBa_2Cu^{2+}_{2+2y}Cu^{3+}_{1-2y}O^{2-}_{7-y}$ but rather $YBa_2Cu^{2+}_2Cu^{1+}_yCu^{3+}_{1-y}O^{2-}_{7-y}$. We can visualize the situation e.g. as if all copper ions in CuO_2 planes remain Cu^{2+} and in chains we have $(1-y)Cu^{3+}$ ions and $y\ Cu^{1+}$ ions. Of course this picture is very crude, actually there is no such strict division between plane and chain sites and there seems to be no Cu^{3+} state either (see below). Each formal "Cu^{3+}" state means just that there is equal number of holes in a system; exact location of these holes is a separate story. Thus in effect we can write down the chemical formula of 123 compound as

$$YBa_2Cu^{2+}_{3-y}Cu^{1+}_y(h)^+_{1-y}O^{2-}_{7-y} \qquad (4)$$

where $(h)^+_{1-y}$ means the concentration of holes (which may go to CuO_2 planes or may occupy copper d-states, oxygen p-states or mixed d-p band states).

We only want to stress that just such an attribution of valence states permits one to understand qualitatively the well known phase diagram of 123 compounds. If for composition between $O_{6.5}$ and O_7 there would be no Cu^{1+} states then $YBa_2Cu_3O_{6.5}$ would contain all Cu ions in Cu^{2+} form and it should be insulating as pure La_2CuO_4. It is known however that this composition is still metallic and superconducting although with a smaller T_c. The transition to an insulating state occurs for still smaller oxygen concentration $\sim O_{6.3-6.4}$. This is explained quite naturally by the formula (4). (Strictly speaking in this case the pure insulator would correspond to O_6 case but similar to $La_{2-y}Sr_xCu_2O_4$ the insulating state can persist up to a certain hole doping). One can see also that T_c in 123 compounds approximately scales with hole concentration (1-y).

Now we discuss in more details where these extra holes responsible for superconductivity are located. One can argue that due to a strong d-d repulsion the $Cu^{3+}(d^8)$ state with extra hole on copper is energetically unfavourable. The idee was put forth that these extra holes should rather be located on oxygen sites /9,10/, $Cu^{3+}O^{2-} \longrightarrow Cu^{2+}O^{-}$. This hypotheses has now strong experimental confirmations /6,7,26/. Therefore we shall discuss in more details the electronic structure of oxygen ions.

4. Oxygen ions: energy level structure

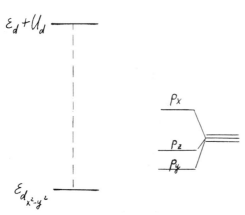

Fig.3. Energy level scheme for d-levels of Cu and p-levels of oxygen in HTSC's (in hole representation). The splitting of p-levels of oxygen is shown (p_x is the orbital directed towards neighbouring Cu ions).

The structure of energy levels of Cu and O which one would obtain taking into account strong d-d interaction is depicted qualitatively in fig.3 (in hole representation). Let us first ignore fine structure of p-levels. In the ground state of La_2CuO_4 in this scheme there is one hole per cell which occupies the lowest $d_{x^2-y^2}$-state. If we suppose that the "extra" holes introduced by doping are located not on Cu but on O-sites we must suppose that (hole) p-levels of oxygen ε_p lie between ε_d and $\varepsilon_d + U_d$.

Usually when discussing the electronic structure of HTSC's one takes into account only oxygen orbitals of p_x, p_y type directed towards neighbouring Cu, see fig.4. These orbitals have large overlap with $d_{x^2-y^2}$-orbitals of copper and they are responsible for antiferromagnetic superexchange interaction and for the formation of hybrid d-p bonding and antibonding bands. In this model extra holes on oxygen are believed to occupy these orbitals (e.g. for oxygen ion A in fig.4 - p_x-orbital). However similar to d-levels of Cu oxygen p-levels in an isolated atom are degenerate. This degeneracy should be also lifted in a crystal due to crystal field effects. It is easy to understand that electron energies of the orbitals directed towards Cu should decrease (respectively hole energies would go up) due to strong attraction of p-electrons to the positive Cu^{2+} ions, see fig.3. From this point of view it is energetically more favourable to put an extra hole not into these states but into the orbitals of p_y, p_z-type looking away from Cu ions e.g. inside Cu_4O_4 square /11,12/.

The splitting of p_x and (p_y,p_z) levels is estimated in Ref.12 as 1-2 eV. If we believe that in the ground state of undoped system all the copper ions are in a Cu^{2+} state (half-filled case, n =1) this would mean that even (p_y,p_z)-hole levels in fig.3 lie above $\mathcal{E}_d$. Then it follows that $\mathcal{E}_{p_x}$ is at least 1-2 eV larger than $\mathcal{E}_{d_{x^2-y^2}}$. In band structure calculations one usually obtains that $\overline{\mathcal{E}}_{d_{x^2-y^2}} \simeq \mathcal{E}_{p_x}$

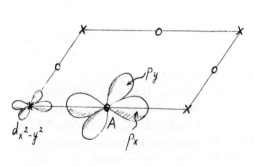

Fig.4. Relevant d and p-orbitals in a basal CuO_2 plane.

However in these calculations the electron-electron interaction is treated in a Hartree-Fock approximation i.e. $\bar{\mathcal{E}}_{d\sigma} = \mathcal{E}_d + U_d \langle n_{d,-\sigma} \rangle = \mathcal{E}_d + \frac{1}{2} U_d \langle n_d \rangle$, and for $U_d \sim 7$ eV and $\langle n_d \rangle \simeq 1$ the coincidence of $\bar{\mathcal{E}}_d$ and $\mathcal{E}_p$ means that for the actual hole levels $\mathcal{E}_{p_x} - \mathcal{E}_{d_{x^2-y^2}} \simeq 3.5$ eV.

The nonbonding p_y-orbitals may form a band by direct overlap with corresponding p-orbitals of neighbouring oxygens. Thus it is possible in principle that p-holes would move in a pure p-band. In this case our systems should be described as a two-component systems with localised d-electrons and delocalized p-electrons - the situation reminiscent of that in mixed valence and heavy fermion compounds. Such a situation was not analysed in details for HTSC but it seems to be rather plausible.

5. p-d hybridization, bands and exchange interaction

In this section we shall discuss in somewhat more details certain properties of Cu-O system especially those connected with p-d hybridization. As d-orbitals of Cu are rather localized one can safely ignore their direct overlap and the essential overlap is the one between d-orbitals and p-orbitals of oxygen and possibly also direct p-p overlap.

We start by discussing the simplest model neglecting the detailed structure of d and p-levels and retaining only $d_{x^2-y^2}$ and $p_{x(y)}$ (p_σ)-orbitals. One can describe this situation by the two-component Hubbard model (we again use here hole representation, i.e. $d^+_{i\sigma}$ and $p^+_{j\sigma}$ are the creation operators for d and p-holes):

$$H = \sum \mathcal{E}_d d^+_{i\sigma} d_{i\sigma} + \mathcal{E}_p p^+_{j\sigma} p_{j\sigma} + U_d d^+_{i\uparrow} d_{i\uparrow} d^+_{i\downarrow} d_{i\downarrow} + U_p p^+_{j\uparrow} p_{j\uparrow} p^+_{j\downarrow} p_{j\downarrow} + (V_{ij} d^+_{i\sigma} p_{j\sigma} + h.c.) \quad (5)$$

We can have different situations depending on the ratio of the parameters U_d, U_p, $\Delta = \mathcal{E}_p - \mathcal{E}_d$ and V /24/. If $\Delta > U_d > V$ i.e. $\mathcal{E}_p > \mathcal{E}_d + U_d$ we may exclude oxygen p-states (which would participate in a charge transfer only virtually) and the situation would be described by the Hubbard model (3), with $t=V^2/\Delta$. For $U_d \gg t$ and $n=1$ the system would be a Mott-Hubbard insulator (region 1 in fig.5).

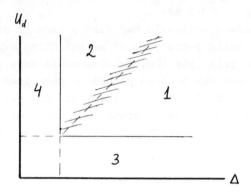

Fig.5. Qualitative phase diagram for a model (5) (after /24/). 1 – Mott-Hubbard insulator. 2 – Charge-transfer insulator. 3 – Metal. 4 – Metal of mixed-valence type.

However if $U_d > \Delta > V$ the situation would be different. In this case doubly-charged d-states would be inaccessible and the relevant excitations would be the charge transfer excitations between d and p-states (between copper and oxygen). Such a situation was shown in fig.3 and it is supposed to correspond to the actual situation in copper metallooxides. For $n=1$ this state (region 2 in fig.5) is again an insulator, of charge-transfer type (the energy gap $E_g \simeq \Delta$) with localized spins on copper coupled by the exchange interaction

$$H = J \sum \vec{S}_i \vec{S}_j \quad , \quad J = t^2 \left(\frac{1}{U_d} + \frac{1}{2\Delta + U_p} \right) \quad , \quad t = V^2/\Delta \quad (6)$$

This exchange interaction is generated by the virtual transitions of a hole from one Cu site to the other via intermediate oxygen (the term with $1/U_d$) or by the transitions of two holes from the neighbouring coppers to the same oxygen ion and back (the term with $1/(2\Delta + U_p)$).

If one takes away the hole from one of the Cu sites (creating Cu^{1+}) such an excitation in this scheme would move through the crystal forming the band with the width $\sim t$ (see fig.6a).

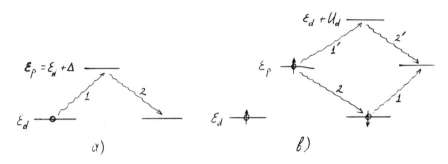

Fig.6. The scheme illustrating the processes leading to the formation of "d" and "p"-bands in La_2CuO_4 (in hole representation; holes are shown by small circles). a) The transfer of "hole deficit" (Cu^{1+} state) from right to the left via oxygen; b) Two possible processes (1,2 and 1',2') leading to the motion of an extra oxygen hole.

If on the other hand we introduce an extra hole then in the situation shown in fig.3 it will occupy p-orbital (here p_x). This extra hole should in this case move through copper and the bandwidth of these p-holes would be

$$t' = V^2 \left(\frac{1}{\Delta} + \frac{1}{(\mathcal{E}_d + U_d) + \mathcal{E}_p} \right) = V^2 \left(\frac{1}{\Delta} + \frac{1}{U_d - \Delta} \right)$$

i.e. of the same order that the bandwidth of d-holes t. (The first term in the expression for t' comes from the transitions 1, 2 in fig.6b and the second one - from 1', 2'). In this case the holes in "d"-band with a bandwidth t spend most time on copper, $|\psi\rangle = \alpha|d\rangle + \beta|p\rangle$, $\alpha^2 \sim 1 - V^2/\Delta^2$ so that an average interaction of electrons in these states would be $U_{eff} \sim U_d \alpha^4 \sim U_d \gg t$ which means that the lower "d-band" should be treated as strongly correlated (Hubbard) one. On the contrary the upper "p"-band will be composed mainly of p-states with only a small admixture of d-states, $|\psi'\rangle = \alpha'|d\rangle + \beta'|p\rangle$, $\alpha'^2 \sim V^2/\Delta^2$, $\beta'^2 \sim 1 - V^2/\Delta^2$ (for $U_d - \Delta > \Delta$), and $U'_{eff} \sim U_d \alpha'^4 + U_p \beta'^4 \sim U_d(V/\Delta)^4 + U_p$. If we ignore U_p the effective "p"-band (hybridized with d-states) could in principle correspond to nearly free-electron band.

The interaction of electrons on oxygen U_p is not known at present. If as is sometimes supposed it is 2-3 eV then this p-band may also have rather strong electron correlations. It is possible however that U_p is small or even negative (see next section).

If in this model d-p hybridization is not small or if some of the energy levels coincide ($\varepsilon_p \sim \varepsilon_d$ or $\varepsilon_p \sim \varepsilon_d + U_d$) the situation becomes much more complicated. In this case it resembles that in mixed valence systems. The main unsolved problem here would be the same: it is not clear how to hybridize the states of completely different nature: strongly correlated (localized) d-electrons (U_d large) and nearly free p-electrons (U_p small).

We want to stress again that whereas in the simple Hubbard model (3) the situation would be symmetric around n=1 under electron-hole interchange (for electron or hole doping of La_2CuO_4) such symmetry is evidently absent for a model (5) (fig.3). Here electron doping would produce Cu^{1+} states and hole doping would give extra holes not on copper but on oxygen. Experimentally we know that such an asymmetry does exist: whereas hole doping ($La^{3+} \longrightarrow Sr^{2+}$) produce metal and superconductor, electron doping ($La^{3+} \longrightarrow Zr^{4+}$, Th^{4+}) gives no such result. This very asymmetry may serve as a strong argument in favour of the model of fig.3 with the "extra" holes on oxygen.

It is instructive also to study the influence of antiferromagnetic ordering on the motion of an extra hole in a model (5). For $\Delta > U_d \gg V$ we reproduce standard situation of the one-band Hubbard model. For Ising spins the hole can move only reversing the localized spins along its trajectory so that it leaves behind the trace of wrong spins /22/. This leads to a strong suppression of hole kinetic energy and gives an effective mechanism by which doping destroys antiferromagnetism.

It is easy to see that the situation would be qualitatively the same in the opposite case $U_d > \Delta \gg V$ (see fig.6b). If the transfer of an oxygen hole occurs only through $\mathcal{E}_d$ and $\mathcal{E}_p$-levels (process (1,2) in fig.6b) and the occupation of doubly-charged d-state $\mathcal{E}_d + U_d$ is forbidden then p-hole would also reverse d-spins on its trajectory leading to similar "confinement". However the situation would be different in the intermediate case $U_d \sim \Delta \gg V$. In this case both the processes (1,2) and (1',2') are possible and it is clear that p-hole may move without disturbing the background antiferromagnetic order of d-spins. Indeed e.g. p-hole with spin up can pass d-site with the same spin by the process (1,2). At the same time it can pass d-site with the antiparallel spin through the doubly-charged d-state (process (1',2')). The only consequence of antiferromagnetic order would be the doubling of the period and corresponding splitting of the p-band due to alternation of transfer integrals V^2/Δ and $V^2/(U_d - \Delta)$; there would be no significant narrowing of the band. Thus we see that in the regions 1 and 2 in fig.5 the situation is similar and one may think that "magnetic" mechanisms of pairing (if efficient at all) would operate in these cases in a similar manner; but strangely enough it seems that in the intermediate case $\Delta \sim U_d$ (cross-hatched region in fig.5) the interplay between hole motion and magnetic ordering would be less important and correspondingly its possible effect on superconductivity would be different.

Returning to the general discussion of the band structure of HTSC we may note that the situation would be somewhat different if the hole on oxygen occupies not the bonding p_x-orbital directed toward Cu but nonbonding orbitals say p_y in fig.4. As we have already noted these orbitals may form a band by themselves with the bandwidth of order 1.5-2 eV. (Such pure p-bands are known to exist in insulators like MgO or ZnO in which p-holes may be produced by the radiation damage). In this case it may turn out that this p-band and hybridized p_x-d band would intersect. Such a two-band picture was invoked sometimes on a purely phenomenological grounds to explain certain properties of HTSC's /13/. The important question would be what is the relative position of these bands and which of them is narrower.

If the hole occupies oxygen p-orbital and is localized (e.g. due to some defects or impurities) it would have rather strong exchange interaction with the copper spins. For p_x-hole this interaction would be antiferromagnetic, with $J_{Cu-O} \sim$ $\sim V^2 \left[1/(U_p+\Delta)+1/(U_d-\Delta) \right]$; for a hole on orthogonal p_y-orbital it will presumably be ferromagnetic, $\sim -V^2 J_H^o/(U_p+\Delta)^2$ where J_H^o is the intraatomic Hund exchange on oxygen. As $J_H \sim 1$ eV both these versions give Cu-O exchange stronger than Cu-Cu one (6). As a result the effective coupling of this pair of Cu spins would become ferromagnetic. This mechanism was invoked by Aharoni et al./12/ to explain the suppression of antiferromagnetic ordering by doping and also as a possible source of superconducting pairing. This is an alternative mechanism to the Anderson RVB picture; it is also based on magnetic interactions but treats it not in a simple Hubbard model but in a more elaborate model with holes on oxygen. The important defect of this approach is that in both cases p-hole can not be treated as localized. On the contrary the hole kinetic energy t is of the same order as exchange interaction J_{Cu-O}. But as we have seen above the consideration of kinetic energy gives qualitatively similar conclusion: the hole motion is suppressed in an antiferromagnetic state but it is completely free in a ferromagnetic one. Thus both these factors destroy antiferromagnetism. But which state would appear instead - ferromagnetic, RVB-like, superconducting - is still an unsolved problem.

6. "Neutral oxygen" and superconductivity

From a chemical point of view not only Cu^{3+} but also O^- state seems to be rather unusual. Such a state is known to exist e.g. in peroxides like H_2O_2 in which formal valence state of oxygen is -1 but two such ions form chemical bond $H{-}O{-}O{-}H$.

In our case also the O^--state at least for localized p-holes would hardly be stable. One should rather expect some tendency to the pairing which may be favourable for superconductivity.

One of the possibilities is that the holes at the neighbouring O-sites would be paired forming so called peroxide ion O_2^{2-}. There exist some indications of the appearance of such a state in HTSC /14,15/. If such an entity would be mobile it may play a role of a superconducting pair; it would be described by the average of the form $\langle p_{j\sigma}^+ p_{j'-\sigma}^+ \rangle$ i.e. it would correspond to a singlet pairing on neighbouring oxygen sites j, j', most probably in a d-wave state.

I would like to point out one extra possibility. It is possible in principle that two holes form a pair not on the neighbouring but on the same oxygen site transforming it into a "neutral" state: $O^- + O^- \to O^{2-} + O^0$. Indeed as contrasted to O^- the states O^{2-} and O^0 are known to exist and are stable. The chemical reaction of this type (so called redox reaction) does really occur at a well known decomposition of hydrogen peroxide

$$H_2O_2 \longrightarrow H_2O + O$$

The hypothesis that similar process may take place in copper metallooxides was put forth in Refs.10,16 on the basis of general "chemical" trends and recalling the well known fact that the oxygen vacancies are formed quite easily in HTSC's. If oxygen exists in these crystals in O^{2-}-form or even in the form O^- it would be difficult to extract it out of the crystal, and only if it can be easily transformed into a neutral form O^0 can it leave the sample.

The tendency to form O^0 state out of two p-holes may be connected with the role of intraatomic exchange and correlation effects and also with the lattice relaxation ("bipolaron" effect). Although at large distances the holes would feel Coulomb repulsion this repulsion may be significantly diminished at short distances due to these effects. Such a situation is known to exist in some systems (Au impurities in Ge, vacancies in Si); in these cases the stable states are those which differ by <u>two</u> electrons (e.g. neutral and doubly-charged vacancy). Similarly some chemical elements prefer certain specific valence states differing by 2: e.g. Bi^{3+} and Bi^{5+} but not Bi^{4+}, Sn^{2+} and Sn^{4+} but not Sn^{3+} etc.

Such a situation may be modelled by the effective p-p attraction in (5) - i.e. "negative U", $U_p < 0$. Another similar model which may correspond to a two-band case is the "negative-U" Anderson lattice

$$H = \sum \varepsilon_k a^+_{k\sigma} a_{k\sigma} + \varepsilon_p p^+_{i\sigma} p_{i\sigma} + \left(V_{ik} a^+_{k\sigma} p_{i\sigma} + h.c. \right) - U_p p^+_{i\uparrow} p_{i\uparrow} p^+_{i\downarrow} p_{i\downarrow} \quad (7)$$

The model (7) seems to be less suitable for a description of copper metallooxides (although it is presumably adequate e.g. for SnTe:Tl superconductors /18/). However one can visualize the situation in which one can get somewhat similar model for 123 compound. Namely it may turn out that due to p-d hybridization the electrons in CuO_2 plane form relatively wide conduction band but the electron states in chains are more localized. Then these chain oxygens may serve as pairing centers (see also /25/). This hypothesis is partially supported by the observation that the oxygen vacancies in 123 compounds are created mostly in chains which means that it is just in chains where the "neutral oxygens" may be formed more easily.

The analysis of both the models (5) and (7) leads to similar conclusions /16/. In one limiting case the "pair" level $E_o = 2\varepsilon_p - U_p$ lies <u>above</u> d-level or Fermi-level (i.e. "neutral oxygen" state is an excited one having energy above that of the ground state). In this case virtual transitions of the electrons from Fermi level to this "pair" level Wp^+p^+aa, $W \sim \frac{V^2}{\varepsilon_p - \varepsilon_F}$ give in a second order an effective attraction of conduction electrons $\sim W^2/(E_o - 2\varepsilon_F) a^+ a^+ aa$ so that the critical temperature of superconducting transition would be

$$T_c \sim \mathcal{E}_F \exp\left\{-\frac{1}{N(0)W^2/(E_0 - 2\mathcal{E}_F)}\right\} \qquad (8)$$

(BCS regime, weak coupling).

In the opposite case of deep pair level E_0 all the electrons (holes) will be bound in pairs which behave as bosons and may bose-condence at $T_c \sim \hbar^2 n^{2/3}/m_B$. In the model (7) the mass of a boson (pair) m_B is $\sim E_0/W^2$ so that in this case

$$T_c \sim W^2/E_0 \qquad (9)$$

(local pair, or Schafroth-Butler-Blatt regime). The optimum value of T_c is reached in an intermediate case when the position of a pair level coincides with the Fermi-level, $E_0 \sim 2\mathcal{E}_F$ (one may call this case double-valence-fluctuation regime). The qualitative form of the dependence of T_c on E_0 (i.e. on the strength of the attraction U_p) is shown in fig.7 (cf. /17/). Similar results may be obtained also in a model (5).

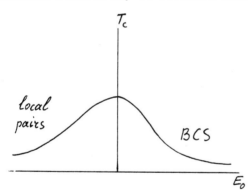

Fig.7. Schematic dependence of T_c on a position of a "pair" level E_0.

The most important question which remains is whether there is really such a tendency to form a "neutral" oxygen. One can check it by the quantum-chemical cluster calculations and possibly by some experiments e.g. by studying photoemission from the oxygen or NMR on ^{17}O. Certain thermochemical results show that there exist in HTSC's up to three different kinds of oxygen; whether it may signify the tendency to form O^0 or has different explanation (e.g. due to different crystallographic positions in 123 lattice) is unclear at present. We only want to stress that for this mechanism to be effective it is not necessary that the actual 0°-level would be stable: even if it lies above ground state but not very high it may lead to an increase of T_c. On the contrary too strong attraction of p-holes would be unfavourable for superconductivity because of the corresponding decrease of T_c, see (9) and fig.7 (not to say of the fact that if neutral oxygen is really formed it would simply go out of the crystal but not promote the pairing).

In the discussion given above we ascribed the tendency to form a "pair" state to the oxygen. It is possible in principle that similar role would be played by other atoms in a crystal e.g. Tl impurities in a superconducting PbTe:Tl /18/ (Tl has a valence 1+ or 3+) or Bi and Tl in respective high-temperature superconductors, including also $BaPb_{1-x}Bi_xO_3$ and $Ba_{1-x}K_xBiO_3$.

References

1. D.Reinen, C.Friebel, Structure and Bonding 1979, 37, 1.
2. I.B.Bersuker, Electronic structure and properties of coordination compounds. Introduction into the theory (in Russian), Leningrad, Chemistry Ed., 1986.
3. M.D.Kaplan, D.I.Khomskii, Problems of high-temperature superconductivity (in Russian), Part I, p.144, Sverdlovsk 1987; Pis'ma v ZhETF (JETP Letters) 1988, 47, 631.
4. W.Weber, Zs. Phys. 1988, B 70, 323.
5. H.P.Geserich, G.Schreiber, J.Geerk, H.C.Li, G.Linker, W.Assmus, W.Weber, preprint 1988.
6. N.Nücker, J.Fink, J.C.Fuggle, P.J.Durham, W.M.Temmermann, preprint 1988; N.Nücker et al., Zs. Phys. B 1987, 67, 9; J.C.Fuggle et al., Phys. Rev. 1988, B 37, 123.
7. A.Fujimori, E.Takayama-Muromachi, Y.Ushida, B.Okai, Phys. Rev. 1987, B 35, 8814.
8. D.Vaknin, S.K.Sinha, D.E.Moncton, D.C.Yohnston, J.M.Newsam,

C.R.Safinya, H.E.King, Jr. Phys. Rev. Lett. 1987, 58, 2802.
9. V.Emery, Phys. Rev. Lett. 1987, 58, 2794.
10. A.K.Zvezdin, D.I.Khomskii, Pis'ma v ZhETF, Prilozhenie 1988, 46, 102 (Sov. Phys. - JETP Lett., Suppl. to vol. 46, s 87, 1988).
11. D.I.Khomskii, A.K.Zvezdin, Physica 1988, C 153-155, 1319.
12. R.Y.Birgenau, M.A.Kastner, A.Aharoni, preprint 1988.
13. V.I.Tsidilkovskii, I.M.Tsidilkovskii, Fiz. Met. i Metalloved. (Phys. of Metals and Metallography) 1988, 65, 83.
14. R.A. de Groot, H.Gutfreund, M.Weger, Solid State Comm. 1987, 63, 451.
15. D.D.Sarma, C.N.Rao, J. Phys. 1987, C 20, L659; D.D.Sarma, K.Streedhar, P.Ganguly, C.N.R.Rao, Phys. Rev. 1987, B 36, 2371.
16. D.I.Khomskii, A.K.Zvezdin, Solid State Comm. 1988, 66, 651.
17. L.N.Bulaevskii, A.A.Sobyanin, D.I.Khomskii, ZhETF 1984, 87, 1490 (Sov. Phys. - JETP 1985, 60, 856).
18. B.A.Volkov, V.V.Tugushev, Pis'ma v ZhETF (JETP Lett.) 1987, 46, 193.
19. G.Aeppli, D.J.Buttrey, preprint (1987).
20. Yu.B.Gaididei, V.M.Loktev, preprint ITP-87-147E, Kiev 1987.
21. P.Day et al., J. Phys. C 1987, C20, L429.
22. L.N.Bulaevskii, E.L.Nagaev, D.I.Khomskii, ZhETF 1968, 54, 1562 (Sov. Phys. - JETP 1968, 27, 836).
23. P.W.Anderson, Varenna Lecture Notes, 1987.
24. J.Zaanen, J.W.Allen, G.A.Sawatzky, Phys. Rev. Lett. 1985, 54, 418; J. Magn. Magn. Mater. 1986, 54-57, 607.
25. I.O.Kulik, A.G.Pedan, Preprint 17-88, FTINT, Khar'kov 1988.
26. Y.Kitaoka, K.Ishida, K.Fujiwara, T.Kondo, K.Asayama, H.Katayama-Yosida, Y.Okabe, T.Takahashi, preprint 1988 (report at Frankfurt Conf. on Crystal Field Effects and Heavy-Fermion Physics).

THE HOLES ON OXYGEN SITES AND BSC PHYSICAL PICTURE IN HIGH-TEMPERATURE SUPERCONDUCTORS

G.M.Eliashberg

L.D.Landau Institute for Theoretical Physics, the Academy of Sciences of the USSR, GSP-1, 117940, Kosygin St.2, Moscow, V-334.

As it follows from the experiment, magnetic and transport cuprate properties are connected with different groups of electron states: magnetic states are localized on the ions Cu^{2+}, whereas current carriers are holes in valence band, genetically related to 2p-oxygen states which overlap weakly with d-copper shell. At a sufficiently large concentration of carriers a magnetic system seems to be in a singlet state and its spinless (singlet) excitation modes are hybridized with phonon modes of the corresponding symmetry. These "dressed" phonons are the main source of interaction between holes, resulting in the formation of the Cooper pairs and superconductivity which in the first approximation can be described in the framework of formalism common for an electron-phonon system. A change of energy of a phonon system at a superconducting transition makes an essential contribution into thermodynamics of a superconducting state. Under certain conditions this transition may become 1-t kind one. A more detailed description requires an exceeding beyond the framework of the Migdal approximation which accuracy in the case of a quasi-two-dimensional hole spectrum is characterized by the parameter $u \sim (\hbar\bar\omega/E_F) \ln E_F/\hbar\bar\omega$ (E_F is the Fermi energy of holes, $\hbar\bar\omega$ is an average phonon energy).

A study of the electron energy-loss spectra [1] and photoelectron spectra [2] has shown that the holes in $La_{2-x}Sr_xCuO_4$ and $YBa_2Cu_3O_7$ belong to 2-p oxygen states, by the way, to those ones, which overlap rather weakly with d-shell of copper ions. These experiments are of great significance to understand a nature of electron properties of cuprates; the main conclusion consists in the fact that in these compounds magnetic properties and current carriers are connected with different groups of electron states. Thus, one should refuse the initial formulated representation that both magnetic and transport properties are determined by strongly hybridized states Cu(3d) and O(2p).

1. Magnetic System

A magnetic system is formed by the spins of ions Cu^{2+}, between which due to hybridization $Cu(3d_{x^2-y^2}^9) - O(2p_{x,y})$ there occurs an exchange interaction with the exchange energy ~0,1 eV (superexchange). These magnetic states are well localized and in the absence of holes on O(2p) (La_2CuO_4, $YBa_2Cu_3O_{7-y}$, $y \gtrsim 0,8$) an antiferromagnetic ordering is realized [3] . A magnetic system evolves strongly when a concentration of holes increases, passing through a complex sequence of states [4]. A detailed discussion of this interesting problem exceeds the limits of the present paper and further we shall only consider system state properties of spins which occur at a sufficiently large concentration of carriers.

First of all, it is easy to see that the exchange energy between the neighbouring copper ions must decrease when a concentration of holes on an oxygen site increases. Really, from electrostatic considerations it follows that the electron transfer energy from O(2p) onto (3d) increases when a negative charge on the oxygen ion decreases and thus, hybridization becomes weaker. One can say that oxygen becomes more aggressive

in this case and the degree of ionicity of the bound Cu-O increases. At the same time it means that localization degree of a magnetic state on Cu will be still increasing. Whence it follows that within the whole interesting interval of hole concentration an adequate desription of magnetic properties can be obtained on the basis of the Heisenberg model (if necessary, taking anisotropy effects into account). At a large concentration of holes the Ruderman-Kittel-Kasuya-Yosida interaction which as known is a long-range and oscillating one , must be included into this model. It seems very truthful that under these conditions a singlet state of spin system suggested by Anderson [5] is realized. In this state a local density of spin is equal to zero and its magnetic nature manifests itself only in the fact that spin excitations (spin fluctuations) must exist. We have no a satisfactory theoretical description of a spectrum of such excitations so far. The Anderson spinon phenomenology [6] by all its attractiveness seems to be in contradiction with the experimentally observed picture. Thus, low temperature linear heat capacity proves to depend strongly on the quality of samples and for the best of them it has too small value to be ascribed to spinons. A study of inelastic neutron scattering [7] does not discover short wave fluctuations with a small energy as compared to J *. A useful information concerning an exchange interaction between spins in a nonmagnetic state can be extracted from the data on a paramagnetic susceptibility. The experiment shows an increase of χ with the growth of con-
_____
* More detailed results of spin fluctuation study were given by G.Aeppli in the report at the workshop in Tbilisi (1988).

centration of carriers [8] : in $YBa_2Cu_3O_{7-y}$ $\chi(y=0.6)=1.8\cdot10^{-4}$, $\chi(y=0)$ = $3.6\cdot10^{-4} cm^3/mol$ at T = 600 K. If one assumes that the whole effect relates to the decrease of J when y decreases then one can estimate J , using the relations of the mean field theory:
$$\chi=(1/3)(g\mu_0)^2 S(S+1) N\cdot\nu/(T+T_0), \quad J=3T_0/(2zS(S+1)); \quad \mathcal{H}_{ex}=-2J(S_1\cdot S_2)$$
(N is the Avogadro number , ν is a number of ions Cu^{2+} per cell, z is a number of the nearest neighbours). If $\nu=2$ (the problem about a state of copper ions in the chains will be discussed in the next section), then at $z=4$ we get:

$$J(y=0.6)\simeq 1.8\cdot 10^3 K, \qquad J(y=0)\simeq 0.75\cdot 10^3 K$$

The first of these numbers is somewhat larger than the one having been obtained from the Raman spectra [9] ($\sim 1,3\ 10^3$K) and via the method of inelastic neutron scattering [10] what is natural at a rather approximate method of estimates. It follows from the results to be discussed further that the Pauli susceptibility is rather not more than 20-30 % of the above mentioned difference of two values χ , and the main reason of the increase of χ is due to the decrease of J when a concentration of carriers increases. At any rate the value J found at $y=0$ gives a lower boundary of the value of the exchange energy between spins in a metallic phase and, thus, a system of spins remains strongly correlated here.

Besides spin fluctuations there must exist nonmagnetic branches of excitation spectrum. Indeed, a system , consisting of N particles with a spin 1/2 has $N!/(N+2/2)!(N/2)!$ of linearly independent states with a total spin equal to zero .

At the given character of interaction one of them is a ground state and moreover, there will exist a number of branches of singlet excitations. In the case of spins localized on the sites of a crystallic lattice a symmetry classification of these branches is possible similarly as it is done for a phonon spectrum and the interval of their dispersion is determined by the order of magnitude by the exchange energy J.

Local fluctuations of charge density connected with fluctuations of a local coordinate symmetry of an electron state (the so-called resonating valence bounds) correspond to these excitations. Thus, a purely electric interaction of singlet excitations with vibrations of a crystal lattice will take place. As known [11] phonon spectra in cuprats contain a number of high-frequency (oxygen) branches and spread up to the energies ~ $10^3 K$. Practically, it coincides with dispersion region of singlet excitations of a spin system which under these conditions will be inevitably hybridized with the phonons of the corresponding symmetry. Really, here we deal with " dressed" phonons, and moreover, the degree of such dressing must be different for vibrations of different types. One of the observable consequences of the mentioned hybridization is an enhancement of anharmonism effects which must occur owing to the fact that a spectrum of a spin system is restricted. Besides, electron and ion degrees of freedom will manifest themselves differently at various ways of studying a phonon spectrum. Hence, neutron scattering is first defined by ions whereas electron energy loss spectra are sensitive to the contribution of an electron component Equally, it refers to intrinsic carriers (holes) as well. We shall consider this problem further.

2. Current Carriers, Resistance and T_c

Experimental and theoretical information concerning "oxygen" hole spectrum and their concentration leaves much to be desired. As is clear now, band calculations made up to present, are based upon a qualitatively erroneous representation about a nature of carriers [12]. Nevertheless, the estimates for some important parameters can be found proceeding from the experimental data. Since the most well-studied system is $YBa_2Cu_3O_7$ (1-2-3) and also $La_{1.85}Sr_{0.15}CuO_4$ then we shall restrict ourselves to the discussion of these very systems though probably some qualitative conclusions hold for other cuprates as well [13].

First of all we shall emphasize that for all high-temperature cuprates almost a two-dimensional character of conductivity is typical and more or less definite information refers first to electron transport phenomena in the so-called cuprate plane CuO_2 whereas a transfer mechanism along the axis c is almost unknown yet. In the best samples 1-2-3 the resistance along the axis is at least 150 times larger than in the plane (a,b) [14].

As has been mentioned above, current carriers in La_2CuO_4 and 1-2-3 occur while forming vacant electron states in 2p-oxygen shell, i.e. when a negative charge of oxygen ion becomes smaller than 2. According to the above said, holes appear first of all in the planes CuO_2. Magnetic and transport states in such a plane are schematically shown in Figs. 1a and 1b, respectively (p-orbitals of oxygen site are denoted by arrows). Since magnetic states are localized, then in the first approximation they should be considered as a source of ion potential. Current carriers are holes in the upper valence band, genetically connected with the states of Fig. 1b. The charge $-2 \cdot e$ on oxygen ions corresponds to the filled band.

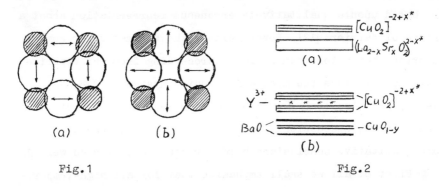

Fig.1   Fig.2

Proceeding from these representations about a nature of carriers, one would try to estimate their concentration from valence considerations. However, due to a layered structure of cuprates there occurs a problem of charge distribution between layers. In La_2CuO_4 the layers CuO_2 and 2LaO interchange and since at a stoichometrical composition this compound is a dielectric, one can expect that the charges of these layers are equal, respectively, $-2e$ and $+2e$ per cell. While introducing Sr an absolute value of these charges decreases (Fig.2a), but beforehand it is far from being evident that $x^* = x$. Judging by the measurements of the Hall constant R_H [15] this equality is approximately fulfilled at $x < 0.2$, but at a further increase of concentration

x^* starts growing more rapidly than x (R_H drops abruptly). In the case $YBa_2Cu_3O_{7-y}$ there is an additional uncertainty connected with the fact that so far it is not quite clear in what valence state a copper ion in chains is at different values of y (Fig.2b). The analysis of the structure parameter behavior while changing an oxygen contents in chains shows that a simple connection between carrier concentration and stoichiometry is also absent here. [16] . Here for a stoichiometrical composition $(y=0)$ n seems to be markedly larger $0,6 \cdot 10^{22}$ cm^{-3} (a hole per elementary cell). It is remarkable that here as y is decreasing there occurs a jump-like transition, accompanied , probably , by an abrupt increase of hole concentration in the planes CuO_2. It shows that not all valence configurations of the 2p-oxygen shell are stable and we cannot control smoothly a concentration of holes on these shells at our own discretion. Here there occurs one more important problem. If in $YBa_2Cu_3O_7$ a hole concentration in the planes CuO_2 is larger than a unit per cell , then what is the role of the redundant electrons? They can stabilize Cu^{1+}-state

in chains but one cannot deny the fact that in this system there exist carriers of both signs. The latter could be one of the reasons of the observed dependence R_H on T [17] . Thus, we have no reliable information concerning a concentration of carriers in 1-2-3. Probably, the value $n \simeq 10^{22}$. cm^{-3} is not too far from the actual one.

Immediately after discovering high-temperature superconductors it was pointed out a resistance increase typical for metals when temperature increases [18]. At the same time other characteristics such as the Hall constant and a thermoelectric power have a more complex behavior [19], [20] (see also reviews [21], [22]). As a result a discussion concerning a metal nature of carriers in these compounds is continued up to now Recently it was discovered that relaxation rate of a nucleus spin ^{17}O in $YBa_2Cu_3O_7$ depends linearly on T in a normal state and has a jump typical for BSC mechanism near a temperature of a superconducting transition [23](Fig.3). This result is a decisive argument in favour of the fact that carriers in cuprates represent a degenerate Fermi-system. They are also an additional confirmation of an oxy - gen nature of holes. A linear temperature behavior was discovered for a relaxation of an electron spin of implanted Fe -ions[24] as well. Finally, there are also the first portraits of the Fermi-sur - face in 1-2-3 [25].

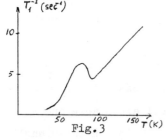
Fig.3

Being sure enough in a metal nature of carriers , still we know very little about their spectrum , even about such characteristics as a density of states averaged over the Fermi-surface. In the simplest case when holes are located near a top of a two-dimensional valence band connected with the $O(2p)$ states of Fig.1b, density of states is defined by a mass m* , so the Fermi energy is directly connected with a concentration of holes. If its own band corresponds to each plane and x^* is a number of holes per cell CuO_2 , then we get

$$E_F = \frac{\pi \hbar^2}{m a^2} \cdot \frac{m}{m^*} \cdot x^* \simeq 1.6 \, \frac{m}{m^*} \, x^* \quad (eV) \tag{1}$$

In $La_{1.85}Sr_{0.15}CuO_4$ where it seems to be $x^* \simeq 0.15$, it rather corresponds to a real picture. Thus, here $E_F \simeq 0.25(m/m^*)eV$. However, in $YBa_2Cu_3O_7$ a concentration of carriers is considerably larger and probably, the Fermi surface has a more complex configuration [25]. In particular, an electron sign of carriers may correspond to its some sections. In this case m* is a very poor spectrum characteristics. The information about m* is usually extracted from the experimental data, referring to plasma frequency, penetration depth of a magnetic field, paramagnetic susceptibility, a jump of heat capacity at a superconducting transition etc.[21],[16],[26], [27]. The most typical estimate obtained in such a way, corresponds to the value $(m^*/m) \sim 5$ and while using data by penetration depth even to $(m^*/m) \sim 10$. A large value m* in combination with the data on resistance results in the conclusion that the interaction of carriers with phonons is too weak and does not play any marked role in the formation of a superconducting state [26]. In the author's opinion this conclusion is premature and causes serious objections. The thing is that due to cuprate specifics all the experimental data numbered above are bad sources of information for a quantative estimate of the electron spectrum parameters. Plasma frequency defined from optic measurements depends on spectrum properties in a wide (~ 1.5 eV) energy interval and moreover, even while using model considerations about a character of spectrum of carriers there remains ambiguity, related to dielectric properties of ion matrix. Paramagnetic susceptibility is considerably defined by a correlated system of spins. Further, probably, one can distinguish a contribution of the Pauli susceptibility, using the measurements of ^{17}O Knight shift, but now there seems to be no such data yet. The use of data of heat capacity also results in a considerable uncertainty under definition m*, as it will be considered below in more detail. As for penetration depth of a magnetic field then here there is a problem general for a whole set of magnetic properties of a superconducting state in cuprates and connected with their strongly marked layered structure. In 1-2-3 compounds pairs of cuprate planes closely located to each other are separated by the layer ~ 8 Å, exceeding a coherence length along the c-axis. Since wave functions of carriers are mainly located in the vicinity of the mentioned planes, then the same must be also referred to a superconducting order parameter. Thus, the latter is not a self-averaged (along the axis c) value. As a result, a penetration depth of a magnetic field must be larger than in an isotropic superconductor, $\delta_L = (m^*c^2/4\pi n e^2)^{1/2}$, with the same volume density of a number of carriers n. Moreover, an additional increase of the observed penetration depth due to fluctuations of phase difference between planes may take place. Here we have a rather important theoretical problem before which clarification the use of data about penetration depth for quantative estimates seems to be unjustified.

Considering the notes made, we shall reexamine the problem about a possible interpretation of cuprate superconductivity on the basis of BSC mechanism with electron-phonon interaction. We shall bear in mind that it can be "dressed" phonons as has been mentioned in section 1. A part of resistivity $\rho(T)$ depending on temperature must have the same origin. Thus, using the observed value T_c and $d\rho/dT$ we shall get restrictions for interaction parameters and carrier spectrum. For the estimates we shall use the formulae, corresponding to the well-known Migdal approximation [28], which applicability is based on a small value of the ratio $\hbar\bar{\omega}/E_F$ where $\bar{\omega}$ is some average phonon frequency. In cuprates a width of a phonon spectrum is $\hbar\omega_{max} \sim 0,1$ eV and as can be seen from (1) E_F is comparatively small, especially, if m^*/m is considerably larger than one. In the next section we shall analyze this problem in more detail.

First consider the conductivity defined by the electron-phonon interaction. We shall calculate it in relaxation time approximation for a purely two-dimensional spectrum and then discuss a correction which should be made if one takes into account a diffference between relaxation time τ and transport time τ_{tr}. In this approximation we get

$$\sigma_{xx} = \frac{e^2}{\pi c_0} \int_0^\infty \frac{d\varepsilon}{2T} \left(ch \frac{\varepsilon}{2T}\right)^{-2} \oint \frac{dl}{v} v_x^2 \tau_p(\varepsilon) \qquad (2)$$

for two planes on the length C_0.
Here $c_c \simeq 13,3$ Å for $La_{1.35}Sr_{0.15}CuO_4$ and $c_0 \simeq 11.7$ Å for $1-2-3$. Integration is performed over a

cross-section of the cylinder Fermi surface by the plane , which is perpendicular to the axis c. In the case of a quadratic spectrum we have

$$\sigma = \frac{ne^2}{m^*} \int_0^\infty \frac{d\varepsilon}{2T} \left(ch\frac{\varepsilon}{2T}\right)^{-2} \tau(\varepsilon) \qquad (2a)$$

In the Migdal approximation (see , for instance [29]) we get

$$\frac{\hbar}{\tau_p(\varepsilon)} = \frac{1}{(2\pi)^3} \int dq_z \oint \frac{d\ell'}{v'} \int_{-\infty}^\infty d\omega \left(cth\frac{\omega}{2T} - th\frac{\omega-\varepsilon}{2T}\right) Im\, K_q^R(\omega) \qquad (3)$$
$$\vec{q} = \{q_z, p-p'\}$$

The function $K_q^R(\omega)$ which includes the Green function of phonons and their interaction with carriers can be represented in the form of a spectral expansion:

$$K_q^R(\omega) = \int_0^\infty d\omega'\, w_q(\omega') \frac{2\omega'}{(\omega')^2 - (\omega+i\delta)^2} \quad , \quad \delta = +0 \qquad (4)$$

In case of purely harmonic phonons $w_q(\omega) = \sum_\nu \alpha_{\nu q}^2 \delta(\omega - \omega_{\nu q})$, where ν is a number of branch of a phonon spectrum. We shall conserve a more general form so that it would be possible to include into consideration excitations of an electron nature hybridized with phonons , having been mentioned in section 1. At temperatures exceeding a phonon spectrum boundary we find

$$\frac{\hbar}{\tau_p} \simeq 2\pi \lambda_p T \quad , \quad \lambda_p = \int_0^\infty \lambda_p(\omega) d\omega \, , \, \lambda_p(\omega) = \frac{1}{(2\pi)^3} \int dq_z \oint \frac{d\ell'}{v'} 2\frac{w_q(\omega)}{\omega} \qquad (5)$$

In a harmonic case $\lambda(\omega)$ does not depend on T and $\rho(T)$ increases linearly. It proves that at not too strict limitations upon $\lambda(\omega)$ a practically linear $\rho(T)$ takes place at rather lower temperatures. To clarify the origin of this phenomenon , let us calculate $\rho(T)$ in the case when there is the only Einstein mode with the frequency ω and the corresponding coupling constant λ :

$$\rho(T) = \frac{m^*}{ne^2} \cdot 2\pi\lambda \cdot f(T), \quad f(T) = \omega \left[sh\frac{\omega}{2T}(1 - \frac{1}{3}th^2\frac{\omega}{2T}) \right]^{-1} \quad (6)$$

In Fig.4 the behavior $f' = df/dT$ is described somewhat schematically. As one can see

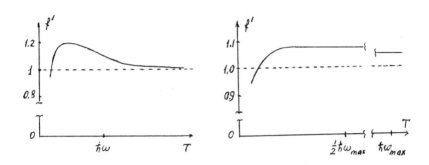

Fig.4   Fig.5

$f'(T)$ tends to a classical limit equal to 1 passing through a maximum at $T < \hbar\omega/2$. We shall emphasize that f' in the maximum is comparatively close to 1. If a phonon spectrum is distributed over a wide frequence interval, then the mentioned maximum transforms into a plateau which extent to higher temperatures T depends on the behavior in the region of large frequences. At considerable variations of $\lambda(\omega)$ the value of f' on the plateau remains practically constant and exceeds 1 per 5-20 % (Fig.5).Thus, in this temperature region we get

$$\rho(T) \simeq -\rho_0 + AT \quad , \quad A = \frac{m^*}{\hbar ne^2} 2\pi\lambda^* \quad , \quad \lambda^*/\lambda \sim 1 \quad (7)$$

A low boundary of the linear section depends on the details of $\lambda(\omega)$ at small frequencies. "A negative zero- T resistance" has been really observed [30]. By the author's opinion it shows that the observed linear dependence $\rho(T)$ is not a result of any exotic mechanism: at sufficiently low temperatures a superlinear behavior $\rho(T)$ must take place (Fig.6).

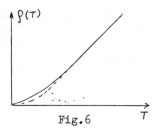
Fig.6

As has been already mentioned, a transport relaxation time $\tilde{\tau}_{tr}$ must enter (2). To define it one should take into account incoming terms in the Boltzmann equation. Their role is especially large in the region of temperatures considerably lower than θ, the Debye temperature, where a scattering of carriers is defined by long-wave acoustic phonons. As known, here $\tilde{\tau}_{tr}^{-1} \sim T^5$. At higher temperatures a contribution of incoming terms has the same scale as at the scattering on impurities. For a quantative analysis the information concerning the interaction of carriers with different branches of a phonon spectrum is quite necessary and we have no it yet. It seems natural that the account of these terms will not change the behavior $\rho(T)$ in the main interval T but the value λ^* must be smaller than the one found above in the relaxation time approximation. When the anomal anisotropy of scattering of carriers in the cuprate plane is absent the estimate $\lambda^* \sim 0.5 \lambda$ seems as reasonable. We shall emphasize that as a result of the account of the incoming terms one should expect an increase of the linear section in $\rho(T)$ to low T, as is shown by a dotted line in Fig.6. Taking the above said into account, we find the relationship

$$\frac{m^*}{\hbar n e^2} 2\pi \lambda = k \cdot \frac{d\rho}{dT} \tag{8}$$

where the coefficient $k \sim 2$ corresponds to the influence of incoming terms. If we use the observed values $d\rho/dT$ for monocrystals $YBa_2Cu_3O_7$, $\sim 0.8 \mu\Omega cm/K$ [14], [30] and $La_{1.85}Sr_{0.15}CuO_4$, $\sim 2\mu\Omega cm/K$ [31], then we get the following restrictions for the parameters of these compounds:

$$\frac{m^*}{m} \cdot \lambda \simeq k \cdot x^* \cdot \begin{cases} 7.0 &, La_{1.85}Sr_{0.15}CuO_4 \\ 3.1 &, YBa_2Cu_3O_7 \end{cases} \tag{9}$$

As above, x^* is a number of holes per cell CuO_2. In $La_{1.85}Sr_{0.15}CuO_4$ probably $x^* \simeq 0.15$ and (9) is really a correlation between λ and m^*. Due to a large concentration Sr monocrystals of this compound have a large zero-T-resistance. In this case the equations for $\rho(T)$ differ from the given above. In the limit when the path length of carriers within the whole temperature interval is mainly defined by scattering on the impurities, i.e. $\tau_{imp}^{-1} \gg \tau_{ph}^{-1}$, one can get a closed expression for $\rho(T)$ depending on temperature: $\rho = \rho_0 + \rho(T)$,

$$\rho(T) = \frac{m^*}{ne^2 \tau_{tr}(T)}; \quad \frac{\hbar}{\tau_{tr}(T)} = \frac{m^*}{(2\pi)^3} \int dq_z \int_0^{2\pi} d\varphi (1-\cos\varphi) \int_0^{\infty} d\omega\, W_q(\omega) \frac{\omega}{2T}\left(sh\frac{\omega}{2T}\right)^{-2} \tag{10}$$

$$\vec{q} = \{q_z, \vec{p}-\vec{p}'\}, \quad \cos\varphi = \frac{(\vec{p}\cdot\vec{p}')}{p \cdot p'}$$

Here the incoming terms are explicitly taken into account.

As has been mentioned in case 1-2-3 there is a considerable uncertainty in the concentration of carriers. Moreover, one can expect a more complicated structure of a spectrum. Here the effective mass defined from different observed parameters can have a different value. Thus, generally speaking, from the data on $\rho(T)$ a lowered value m^* is obtained. Nevertheless, on the ground of (9), probably one can conclude that large values $m^* \sim 5 \div 10$ are inconsistent with the value $\lambda > 1$. We shall emphasize that in the paper by the author [32] incoming terms were not taken into account and moreover, the value n was used which is most likely a lowered one. It resulted in more strict limitation for λ and m^* as compared to (9).

The purpose of a further discussion is as follows. We shall try to show that the experimental information does not contradict to the assumption that superconductivity in cuprates is defined by the interaction which can be called an enhanced electron-phonon interaction. The sense of this notion has been partially explained above. Now let us consider this problem in more detail.

First of all the experiment shows that no doubt, phonons play a certain role while forming the Cooper pairs: an isotope effect [33] exists, there is a pronounced phonon structure in volt-ampere characteristics of the tunnel junctions. In this respect the results [34] are the most instructive. Moreover, the presence in a phonon spectrum of high-frequence modes related to oxygen vibrations and an "oxygen" nature of carriers determined now create the appropriate conditions to realize a phonon mechanism. However, a specifics of cuprates is not exhausted by it. Here a whole number of modes of an electron origin soft in the atomic scale of energies and related to the presence of magnetic ions Cu^{2+} in the lattice must exist.

Rather more stiff modes originate from the orbital momentum $\ell = 2$ of a free copper ion. Due to a low symmetry of a crystallographic positions of copper an orbital degeneration is removed and a spherical polynomial x^2-y^2 corresponds to the ground state. Weber pointed out that in the crystal lattice, containing such ions, there may exist spectrum branches, similar to phonons, but related to the fluctuations of the quadrupole momentum of Cu-ions [35]. According to his estimate the energy of such vibrations is 0,3-0,5 eV what is several times larger than the width of the phonon spectrum. Thus, they do not make a contribution into the resistance, but can play a role in the interaction between holes. Optic measurements discover a transition in this region of the spectrum [36]. On the other side, a calculation shows that the state $3d^9_{x^2-y^2}$ is separated from others by the interval $\sim 2eV$ [37]. Measurements of the nuclear quadrupole resonance (NQR) give an important information about a state of Cu-ions in cuprates. In particular, it turns out that in $YBa_2Cu_3O_{7-y}$ the value of the quadrupole splitting for Cu(2) (in the cuprate plane) increases as y decreases [38]. It shows that a positive charge, localized on Cu-ion is maximal at $y = 0$. It means it its turn, that hybridization Cu(3d) - - O(2p) decreases together with y, as has been mentioned in section 1. Softerning of d-d transitions must be a result of it.

Another series of electron modes is connected with the spin Cu^{2+}. As has been mentioned briefly in section 1, there must exist both excitations having a spin and singlet ones. Both of them have dispersion defined by the exchange energy of interaction J between copper ions, $\sim 0{,}1$ eV. A Straightforward information about a spectrum of spin fluctuations in metallic phases of cuprates is absent yet. Shortwave fluctuations with the energy $\sim 0{,}1$ eV seem to make the main contribution into a total density of states. This region is difficult for neutron studies. As for singlet excitations of a spin system, as has been mentioned above, a considerable part of their dispersion region overlaps with a phonon spectrum. Due to a purely electric nature of singlet excitations in this region their hybridization with phonons must take place. In particular, it refers to those lattice vibrations which are accompanied by a change of bounds Cu-O. However, near the edge of a phonon spectrum /by 10^3 K/ and beyond its limits singlet excitations separate themselves from a motion of ions, and thus, here their direct observation is possible. With this respect one should pay much attention to several experimental results. The tunnelling measurements discovered that in 1-2-3 besides typical phonon maxima there exists a considerable singularity, corresponding to excitations with the energy $(1 - 1{,}3) \cdot 10^3$ K [34]. Electron energy-loss spectra in the region of small energies have a maximum in the same interval [38]. Strongly marked singularities at the same energies [39] have microjunction spectra of a thetragonal phase 1-2-3. Absorption in this spectrun region was pointed out earlier [40]. We shall emphasize that tunnel characteristics $La_{1.85}Sr_{0.15}CuO_4$

do not have a mentioned singularity [34] . It is very probable that here soft electron modes described above manifest themselves , though one cannot exclude another interpretation either, which was given, for instance, by Tachiki and Takahashi [41] . At any rate, I agree with these authors that , probably, here we have one of the key elements of the mechanism of high-temperature superconductivity.

Now there are little doubts that cuprate superconductivity in its main features corresponds to the Bardeen-Cooper-Schrieffer theory. Many objections which existed before were removed by the measurements of temperature dependence of relaxation rate of the nuclear spin ^{17}O [23] (Fig.3). Thus, to describe superconductivity in these compounds one can use the existing theoretical scheme [29,42]. In the next section we shall discuss in what extent it needs precision. In the framework of this scheme a transition temperature is defined by the following equation

$$\chi_p(n) = \left[\frac{1}{(2n+1)(1+Z_p(n))}\right]^{1/2} \sum_{n'\geq 0} \left[\frac{1}{(2n'+1)(1+Z_{p'}(n'))}\right]^{1/2} \overline{K}_{p,p'}(n,n') \chi_{p'}(n') \Big\rangle_{p'} \quad (11)$$

where $\langle \cdots \rangle_{p'} = \oint \frac{d\ell'}{v'} (\cdots) / \oint \frac{d\ell}{v}$

$$Z_p(n) = \frac{1}{2n+1} \sum_{n'\geq 0} \langle K_{p,p'}(n,n') \rangle_{p'}$$

$$\overline{K}_{p,p'}(n,n') = \int_0^\infty d\omega \, \overline{\lambda}_{p,p'}(\omega) D^{(+)}(\omega;n,n')$$

$$K_{p,p'}(n,n') = \int_0^\infty d\omega \, \lambda_{p,p'}(\omega) D^{(-)}(\omega;n,n')$$

$$D^{(\pm)}(\omega;n,n') = s^2 \left(\frac{1}{s^2+(n-n')^2} \pm \frac{1}{s^2+(n+n'+1)^2}\right) ; \quad s = \omega/2\pi T_c$$

We have used a representation of the type (4) for the Green function describing an interaction between carriers, so $\bar{\lambda}_{P,P'}(\omega)$ is connected with a spectral density $\overline{W}_q(\omega)$:

$$\bar{\lambda}_{P,P'}(\omega) = \frac{1}{(2\pi)^3} \int dq_z \, 2 \, \frac{W_q(\omega)}{\omega} \; ; \; \bar{\lambda}_P(\omega) = \langle \bar{\lambda}_{P,P'}(\omega) \rangle_{P'} \qquad (12)$$

In case of a two-dimensional spectrum, which is isotropic in the plane CuO_2, $\bar{\lambda}_P(\omega)$ does not depend on the direction on P on the Fermi surface, and in case of a purely phonon interaction the introduced definition (12) coincides completely with the one which is usually used

$$\bar{\lambda}(\omega) = 2 \, \frac{\alpha^2(\omega) F(\omega)}{\omega} \; , \; \bar{\lambda} = \int_0^\infty d\omega \, \bar{\lambda}(\omega) \qquad (13)$$

According to the discussion made we shall also include the contribution of the above mentioned branches of a spectrum of an electron origin into $\overline{W}_q(\omega)$. In this case a separation of this value for an interaction function and a density of states is less unambiguous. It also refers to a phonon region of spectrum, since phonons / to a larger or smaller extent for different branches/ seem to be hybridized with the mentioned singlet excitations. This hybridization may result in a certain enhancement of interaction of carriers with phonons.

Now consider the difference between $\overline{W}_q$ and W_q, entering the equation for resistivity. This difference appears because of the role which spin fluctuations might play.

As known, for s-pairing exchange interaction of carriers with spin fluctuations gives a negative ontribution into the kernel of equation (11). If $\lambda_{p,p'}^{(ex)}(\omega)$ is the corresponding interaction function, then

$$\bar{\lambda}_{p,p'}(\omega) = \lambda_{p,p'}(\omega) - 2\lambda_{p,p'}^{(ex)}(\omega) \qquad (14)$$

So far we know very little about spin fluctuations in the metallic phase of cuprates and especially, about their interaction with current carriers. There is just a small volume of qualitative experimental information. A rather interesting observation consists in the following. Spin lattice relaxation rate of a nuclear spin ^{63}Cu, T_1^{-1} measured for copper sites in the plane (Cu(2)) for $YBa_2Cu_3O_y$ increases by $\sim 10^3$ times when y increases from 6.52 ($T_c \simeq$ 60K) up to 6.91 ($T_c \simeq$ 90 K) [43]. We shall point out that T_1^{-1} in $La_{1.85}Sr_{0.15}CuO_4$ has the same order of magnitude as in $YBa_2Cu_3O_{6.52}$ [44]. From the viewpoint of the physical picture discussed in this paper, one can give the following qualitative explanation of the mentioned effect. In the absense of interaction with carriers in spin system Cu^{2+}, being in a rather strong correlated nonmagnetic singlet state, the intensity of low-frequency spin fluctuations is rather small. Since these very fluctuations are responsible for nuclear spin relaxation (Cu), then the rate of the latter will be small in this case. Due to exchange interaction with carriers high frequency spin fluctuation might decay into a particle-hole pair in the vicinity of the Fermi surface. Thus, there opens a relaxation channel of a nuclear spin through the intermediate excited state of the system of spins Cu^{2+} with the energy $\sim J$. Further, it is natural that the main part of the density of states of this system is related to short-range wave fluctuations , which wave vector has the order of magnitude π/a (a is a constant of the lattice CuO_2). On the other side , a maximal wave vector of particle-hole pair is $2 p_F$. In $La_{1.85}Sr_{0.15}CuO_4$ and $YBa_2Cu_3O_{6.5}$ a concentration of holes is small and p_F is considerably smaller than π/a . Thus, for the main part of the phase volume a decay of spin fluctuations is forbidden. In $YBa_2Cu_3O_7$ a concentration of carriers is considerably larger , what results in the increase of the decay rate. At the transition into a superconducting state in the spectrum of carriers there opens a gap and, hence, T_1^{-1} decreases , as the experiment shows .

These considerations refer to the interaction of carriers with singlet excitations as well. Proceeding from the fact that it is not related to the exchange , one can assume that a total contribution into interaction kernel of equation (11) has a positive sign. According to the above said , one should expect that this contribution in the case $YBa_2Cu_3O_7$ will be considerably larger than in $La_{1.85}Sr_{0.15}CuO_4$. This conclusion seems to be confirmed by the results of tunnel measurements [34] , having been mentioned above.

To take the Coulomb repulsion in (11) into account, the Coulomb dimensionless pseudopotential $-\mu^*$ is usually introduced. By the same way one can take into account a contribution of electron transitions with the energies $\gtrsim 0.3 - 0.5\ eV$, d-d transitions , in particular , since E_F seems to have the va-

-lue compared with the indicated one. If really these transitions play some role, then, as a matter of fact, it means that μ^* becomes smaller by the absolute value or it even changes the sign. An experiment does not allows one to get any reliable estimate of the effective μ^* so far. However, one can say definitely that if $\mu^* < 0$ and $|\mu^*| \geq 0.2$ then it provides $T_c \sim 100$ K and thus, other interactions do not play any role in superconductivity. Many facts, some of them being mentioned above, contradict to such an assumption. In the opposite case when μ^* does not play any role, via (11) one can estimate the interaction constant $\bar{\lambda} = \int \bar{\lambda}(\omega) d\omega$, which provides the necessary value T_c. The result depends on frequency dependence $\lambda(\omega)$, but taking linearity condition for $\rho(T)$ into account, one can get

$$\bar{\lambda} \sim 0.8 \div 1.2 \quad (La_{1.85} Sr_{0.15} CuO_4); \quad \bar{\lambda} \sim 1.5 \div 2.3 \quad (YBa_2Cu_3O_7)$$

A low boundary corresponds to a large contribution of frequency region in the vicinity of the phonon spectrum boundary $\hbar\omega_{max}$. According to (14) λ is somewhat larger than $\bar{\lambda}$. Here a quantative information is absent yet.

Together with (9) it results in the values of m^*, markedly smaller than those obtained in the above mentioned papers. An additional experimental information is necessary in order to estimate a possible contribution of the negative pseudopotential.

3. Phonon contribution into thermodynamics of a superconducting state. Corrections to the Migdal approximation.

An important source of information about density of states on the Fermi level is a heat capacity jump at $T=T_c$, ΔC. For 1-2-3 compounds its value is in the limits 4,5-6 J/mol K [45]. If one calculates electron heat capacity of a normal state (unfortunately, it is practically impossible to measure it), then for a two-dimensional spectrum with one valley per a cuprate plane we find for the molar volume the following:

$$C = \frac{\pi}{3} \frac{m^*}{m} \cdot \frac{Na^2 m}{\hbar^2} (1 + \lambda(T)) \cdot T \qquad (15)$$

where a^2 is a square of the cell CuO_2. The function is connected with a real part of the self- energy of electrons [42], $\lambda(0) = \lambda$, but $\lambda(T_c) \simeq (0.7 \div 0.8)\lambda$ depending on the form of the spectral density $\lambda(\omega)$. For two planes per cell :

$$C \simeq 2.9 \frac{m^*}{m} (1 + \lambda(T)) \cdot T [K] \; mJ/mol K \qquad (16)$$

We emphasize that for a two-dimensional parabolic band C does not depend on hole concentration. Using the experimental value ΔC, we shall find

$$\frac{\Delta C}{C} \simeq (17 \div 23) \frac{m}{m^*} \cdot \frac{1}{1 + \lambda(T)} \qquad (17)$$

As known, in BSC theory this ratio is equal to 1,43 .

A maximal theoretic value, found for a rather exotic phonon spectrum is equal to 3,73. For a more realistic spectrum it is ~2 [46]. This relation together with (9) could be used to define m^* and λ. However, here one should be careful: if there are two or more groups of carriers with different masses, then "heavy" carriers make an essential contribution into heat capacity, whereas conductivity is mainly defined by a light component.

There exists one more effect which, at any rate, makes probably an essential contribution into heat capacity jump. The thing is that while passing into a superconducting state, a free energy of a phonon system is also changed. In the usual superconductors this contribution is small and constitutes $(\theta/E_F)^2$ of all energy of phonons which is small in itself in the region $T \ll \theta$. In cuprates effect enhancement occurs by the following reasons. First of all, multimode phonon system at $T \gtrsim T_c$ possesses a large heat capacity. Thus, in 1-2-3 $C_{ph} \sim 120$ J /mol K, what is 2 orders larger than an electronic one. Moreover, there occurs an increase of the contribution of phonons to ΔC owing to a quasi-two-dimensional hole spectrum. As will be explained below, in the Ginsburg-Landau equation there appears an additional term which order of magnitude is clear from the following expression:

$$a\psi^2 + b(1-u)\psi^3 = 0 \quad, \quad u \sim \frac{\bar{\omega}}{E_F} \ln \frac{E_F}{\bar{\omega}} \tag{18}$$

The Fermi energy in 1-2-3 is comparatively small (~ 0.6 eV, $m^* \simeq 2.5 m$) whereas a characteristic frequence of phonons $\bar{\omega}$ can be put equal to $0.5 \cdot 10^3$ K.

For our purpose it is convenient to use the expression for a thermodynamic potential Ω which has been obtained via Luttinger and Ward method [42]:

$$\Omega = -2T \sum_P \{\tfrac{1}{2}\ln\varphi(P) + \Sigma_1(P)G(P) - \Sigma_2(P)F(P)\} + \qquad (19)$$
$$+ \tfrac{1}{2}T \sum_Q \{\ln D^{-1}(Q) + \Pi(Q)D(Q)\} +$$
$$+ T^2 \sum_{P,P'} \alpha^2_{P-P'} [G(P)G(P') - F(P)F(P')] D(P-P').$$

Here the following notations are used:

$$\varphi(P) = [\tilde{\xi} - i\varepsilon_n - \Sigma_1(P)][\tilde{\xi} + i\varepsilon_n - \Sigma_1(-P)] + [\Sigma_2(P)]^2,$$
$$G(P) = [\tilde{\xi} + i\varepsilon_n - \Sigma_1(-P)]/\varphi(P), \quad F(P) = \Sigma_2(P)/\varphi(P),$$
$$D^{-1}(Q) = D_0^{-1}(Q) - \Pi(Q), \quad D_0(Q) = 2\omega_q/(\omega_m^2 + \omega_q^2),$$
$$P = \{i\varepsilon_n, \vec{p}\}, \quad Q = \{i\omega_m, \vec{q}\}, \quad \varepsilon_n = (2n+1)\pi T, \quad \omega_m = 2m\pi T$$

The first variations Ω in Σ_1, Σ_2 and Π vanish provided, these values satisfy the equations corresponding to the Migdal approximation

$$\Sigma_1(P) = T \sum_P G(P') D(P-P') \alpha^2_{P-P'} \qquad (20)$$
$$\Sigma_2(P) = T \sum_P F(P') D(P-P') \alpha^2_{P-P'}$$
$$\Pi(Q) = -2T \sum_P [G(P)G(P') - F(P)F(P')] \alpha^2_q, \quad (P' = P-Q)$$

(the definitions of the Green functions differ by the sign from [42]). Expansion Ω in Σ_2 starts from the terms of the fourth order provided Σ_2 is determined from (20). Let's $\delta\Pi$ is part Π, containing Σ_2^2, include Π_0 into

D_o. Then in Ω while taking this contribution into account there appears an additional term $\sim (\delta \Pi)^2$:

$$\delta\Omega = \frac{1}{4} \sum_Q [\delta \Pi(Q) D_o(Q)]^2 \qquad (21)$$

If one introduces the notation $\Sigma_2(\varepsilon) = \psi \tilde{\sigma}_2(\varepsilon)$, where $\tilde{\sigma}_2(\varepsilon)$ is a somewhat normalized function, then we get:

$$\Omega_s - \Omega_n = -\frac{1}{2} \beta (1-u) \psi^4$$

where u corresponds exactly to the additional term in (18). As a result for the energy of a superconducting state we shall get $(q = \alpha(T-T_c))$:

$$\Omega_s - \Omega_n = -\frac{\alpha^2}{2\beta} \cdot \frac{1}{1-u} (T_c - T)^2 ; \quad u = \beta \frac{\bar{\omega}}{E_F} \ln \frac{E_F}{\bar{\omega}} \qquad (22)$$

We shall point out that a positive u is defined by the fact that Ω is minimal relative to the variations $\delta \Pi$. Thus, a phonon contribution results in an additional energy decrease of a superconducting state and, simultaneously, increases a heat capacity jump. The preliminary rough estimates show that β can be markedly larger than a unit, and the effect is considerable even at $m^* \sim m$, when E_F is large.

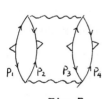

Fig. 7

Fig. 8

In Fig. 7 a diagram is described, corresponding to (21), where wavy and solid lines are phonon and electron Green functions, Σ_2 corresponds to triangles. There are several diagrams of the same type. The main contribution occurs from the regions where one of the following momenta : $q_1 = p_1 - p_2$, $q_2 = p_2 - p_3$, $q_3 = p_1 + p_3$ is small, an essential integration region being $q_i \sim \bar{\omega}/v_F$. However, it is from these regions that there occur corrections for the Migdal approximation. For instance, a diagram of Fig.8 gives a contribution of the same order, moreover, it has another sign. A compensation is only partial, since in the loop of Fig.7 there is a summing over spins. Besides, for each of the numbered channels q_i there exists a sequence of ladder diagrams giving a contribution of the same scale. They do not contain a logarithm, but are essential while defining $\bar{\omega}$ in (18). Diagrams of the type of Fig. 9 are especially interesting : they correspond to the so-called Cooper ladder and in the small vicinity of T_c contain a fluctuation contribution. A more detailed analysis will be published elsewhere.

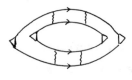

Fig. 9

We shall emphasize one more qualitative result. Two loops in Fig. 7 can refer to both one and the same and different cuprate planes. It is clear, that at least for two neighbouring planes in I-2-3 compounds this contribution will be of the same scale. There is no such an effect in the diagram of Fig.8 and thus, it gives a contribution which is approximately 4 times smaller than of

Fig. 7. Thus, there occurs a peculiar interaction between planes even in the absence of tunnellling which lowers additionally the energy of a superconducting state. There arises a question : can a coefficient at ψ^4 in (18) change a sign? In this case we would deal with the transition of the first kind. Here a careful analysis is needed. Preliminary calculations show that $u \sim 1$ at $m^* \sim 2.5 m$. At any rate, it should be taken into account while using Δc for different estimates.

I would like to thank A.Golubov, L.Gor'kov , D.Khmelnitskii, L.Levitov , E.Rashba , V.Timofeev and I.Schegolev for valuable discussions. I am also grateful to E.Abrahams, G.Aeppli, B.Batlogg , S.Doniach , Z.Fisk, A.Malozemoff , D.Pines and D.Scalapino for many useful conversations during the American-Soviet workshop (Tbilisi , 1988).

References
1. N.Nucker, J.Fink, J.C.Fuggle, P.J.Durham , W.M.Temmerman Phys.Rev. B37 (1988), 5158.
2. J.C.Fuggle, P.J.W.Weijs, R.Schoorl et al, Phys.Rev. B37, (1988), 123; T.Takahashi , F.Maeda, H.Arai et al; Phys.Rev. B36 (1987) 5686; T.Takahashi, F.Maeda, H.Katayama-Yoshida et al, Phys.Rev. B37 (1988), 9798.
3. D.Vaknin , S.Sinha , D.Monkton et al , Phys.Rev. Lett. 58 (1987), 2802.
4. T.Fujita et al , Jap.Journ.Appl.Phys., Suppl.26-3, (1987), 1041.
5. P.W.Anderson , Mater.Res.Bull. 8 , (1973), 153.
6. P.W.Anderson , Science 235 (1987) 1196.
7. T.Brückel, H.Capellmann , W.Just et al, Europhys.Lett., 4, (1987), 1189.
8. D.C.Johnston , S.K.Sinha , A.J.Jacobson , J.M.Newsam, Proc.

Internat.Conf., Interlaken , (1988), p.572.
9. K.B.Lyons, P.A.Fleury, L.S.Schneemeyer , J.V.Waszczak . Phys.Rev.Lett. 60 (1988),732.
10. G.Shirane, Y.Endoh, R.J.Birgeneau et al , Phys.Rev.Lett., 59 (1987), 1613.
11. V.D.Kulakovskii, O.V.Misochko, V.B.Timofeev et al, Pis'ma JETP 46 (1987), 460 ; A.V.Bazenov, L.V.Gasparov , V.D.Kulakovskii et al, Pis'ma JETP 47 (1988), 162; B.Renker, F. Gompf, E.Gering et al, Z.Phys.B67 (1987), 15 ; P.Strobel , P.Monceau, J.L.Tholence et al, Proc.Internat.Conf, Interlaken , 1988.
12. L.F.Mattheiss Phys.Rev.Lett 58 (1987), 1028; J.Yu, S.Massida, A.J.Freeman , D.D.Koelling Phys.Lett. A122 (1987) , 203.
13. C.Michel, M.Hervieu, M.M.Borel et al, Z.Phys.B68 (1987), 421; H.Maeda, Y.Tanaka, M.Fukutomi , T.Asano, J.Appl.Phys. Lett., submitted ; Z.Z.Sheng, A.M.Hermann, Nature 332 (1988), 55 and 138.
14. Yu.A.Ossipyan, V.B.Timofeev, I.F.Schegolev , Proc.Intern. Conf., Interlaken 1988, p.1133.
15. M.Petraviĉ , E.Tutiš, A.Hamziĉ , L.Forro, Sol.St.Comm. 65 , 1988 , 573.
16. D.R.Harshman, L.F.Schneemeyer, J.V.Waszczak , G.Aeppli , R.J.Cava, B.Batlogg et al; Preprint , 1988.
17. Y.Iye, T.Tamegni et al , Physica C 153-155 (1988) ,26.
18. J.G.Bednorz, K.A.Müller, Z.Phys.(1986), 189; C.W.Chu, P.H. Hor, R.L.Meng, L.Gao, Z.J.Huang, Y.Q.Wang Phys.Rev.Lett. 58 (1987), 405.
19. Uchida, H.Takagi, H.Ishii et al, Jap.J.Appl.Phys. 26 (1987) L440; R.S.Kwok, S.E.Brown, J.D.Thompson , Z.Fisk, G.Gruner Physica B148 (1987), 346.
20. R.C.Yu, M.J. Naughton , X.Yan et al ;Phys.Rev. B37 (1988), 7963.
21. L.P.Gor'kov, N.B.Kopnin Usp.Fiz.Nauk 156 (1988), 117.
22. Y.Iye, preprint 1988.
23. Y.Kitaoka, R.Ishida, K.Asayama, H.Katayama-Yoshida ,Y.Oka-

ba, T.Takahashi, Nature (1988), submitted.
24. D.Riegel, S.N.Nishra et al , Phys.Lett, $\underline{A131}$, (1988), 533.
25. L.C.Smedskjaer, J.Z.Liu, R.Bedenek , D.G.Legnini, D.J.Lam, M.D.Stahulak , prerint , 1988 ; M.Peter, L.Hoffmann, A.A. Manual Proc.Internat.Conf., Interlaken , 1988.
26. M.Gurvich, A.T.Fiory , L.S.Schneemeyer et al .Proc.Intern. Conf., Interlaken 1988, p.1369.
27. V.Z.Kresin , S.A.Wolf Journ.of Supercond. $\underline{1}$ (1988), 143.
28. A.B.Migdal , JETP 34 (1958), 1438.
29. G.M.Eliashberg , JETP $\underline{39}$, (1960), 1437.
30. A.Kapitulnik, Proc.Internat.Conf., Interlaken 1988 , p.520.
31. M.Suzuki, T.Murakami, Jap.Journ.Appl.Phys. $\underline{26}$ (1987), L.524.
32. G.M.Eliashberg , Pis'ma JETP $\underline{48}$ (1988) 275.
33. B.Batlogg , G.Kourouklis, W.Weber et al , Phys.Rev.Lett . $\underline{59}$ (1987), 912.
34. L.N.Bulaevskii, O.V.Dolgov, L.P.Kazakov, S.N.Maksimovskii, M.O.Ptitsyn, V.A.Stepanov, S.I.Vedeneev; preprint 1988.
35. W.Weber , prerint 1988.
36. K.Kamaras, C.D.Porter, M.G.Doss et al, Phys.Rev.Lett. $\underline{59}$ (1987), 919.
37. C.H.Pennington, D.J.Durand, C.P.Slichter et al, preprint 1988.
38. Y.Fukuda, M.Nagoshi, Y.Namba, Y.Syono, M.Tachiki (see Ref. in the paper Tachiki and Takahashi [41]).
39. I.K.Janson, L.F.Rybalchenko et al , Fiz.Niz.Temp. $\underline{14}$ (1988) 886.
40. B.I.Verkin, V.M.Dmitrijev Fiz.Niz.Temp. $\underline{13}$ (1987) 849.
41. M.Tachiki , S.Takahashi , Phys.Rev. $\underline{B38}$ (1988), 218.
42. G.M.Eliashberg , JETP $\underline{38}$ (1960), 966; $\underline{43}$ (1962), 1105.
43. H.Yasuoka , T.Shimizu , T.Imai , S.Sasaki preprint 1988.
44. B.A.Alexashin, S.V.Verhovskii et al.Fiz.Met. i Met.$\underline{64}$ (1987), 392.
45. M.Ishikawa, Y.Makazawa, T.Takabatake, Techn.Rep.ISSP, Ser.A, N 1907 (1988); K.Kadowaki, M.Van Sprang, Y.K.Huang et al , Physica $\underline{B148}$, 442 , 1987.
46. J.Blezius, J.P.Carbotte Phys.Rev. $\underline{B36}$ (1987), 3622.

POLARON THEORY OF HIGH-Tc SUPERCONDUCTORS

A.S.Alexandrov

Moscow Physical Engineering Institute, Moscow, USSR

It is shown that the consistent theory of strong electron-phonon interaction, taking into account the small polaron and small bipolaron formation describes satisfactorily the main properties of high-Tc metallic oxides.

The discovery by Müller and Bednorz[1] and by Chu et al.[2] of high-Tc superconductors gave a powerful impetus to the emergence of numerous theories of high-Tc superconductivity. In this connection two most important questions arise: firstly, what type of theory, namely mean-field BSC-like or local-pair i.e. bipolaron [3,4] is relevant? and secondary, what type of interaction namely Coulomb[5] or electron-phonon correlation is responsible for pair formation?

In this paper we shall discuss a wide range of reliable experimental data which seem to prove the applicability to LBSO, LSCO, YBCO and other metallic oxides of our many-polaron theory of strong-coupling superconductors [3,4].

As we have previously noted[6] the strong-coupling condition

$$\lambda \geq 1 \qquad (1)$$

is practically identical to the one for small polaron formation [7]

$$\frac{f(z)g^2\omega}{D} \geq 1, \qquad (2)$$

under which a strong renormalization of the electron spectrum occurs resulting in an exponential reduction of the initial electronic bandwidth 2D to an extremely narrow polaronic band with a half width:

$$W = (D) \exp(-g^2) \qquad (3)$$

here $f(z) \simeq (2z)^{1/2}$, z is a nearest neighbours' number, g^2 and ω are a dimensionless electron-phonon interaction constant and a characteristic phonon frequency determined by the ordinary Frölich Hamiltonian in momentum or site representation:

$$H_{e-ph} = \sum_{\bar{q}\bar{k}} U(\bar{q}) \, C^+_{\bar{k}+\bar{q}} \, C_{\bar{k}} \, d_{\bar{q}} + h.c.$$

$$= \sum_{\bar{q}\bar{n}} U(\bar{q}) e^{i\bar{q}\bar{n}} \, C^+_{\bar{n}} \, C_{\bar{n}} \, d_{\bar{q}} + h.c. \, ,$$

$$g^2 = \sum_{\bar{q}} \omega_{\bar{q}}^{-2} \, cth \, (\omega_{\bar{q}}/2T) \, U^2(\bar{q}) \, (1-\cos(\bar{q}\bar{n})), \quad (4)$$

$$\omega = g^{-2} \sum_{\bar{q}} \omega_{\bar{q}}^{-1} \, U^2(\bar{q})$$

$C_{\bar{q}}$ and $d_{\bar{q}}$ are the electron (hole) and phonon operators respectively, $\omega_{\bar{q}}$ is the phonon dispersion, $\bar{n}$ labels the site.

In such a way the well-known Migdal theorem breaks down in the strong-coupling limit (1)./8/ Instead the system is in the so-called antiadibatic regime

$$E_F \lesssim W \lesssim \omega \quad (5)$$

where E_F is the renormalized Fermi energy.

As the many-polaron theory shows the taking into account of the polaron effect (3) qualitatively changes the nature of the superconducting state: in the intermediate coupling region $\lambda \simeq 1$ the ordinary BSC is replaced by polaron /4/ and bipolaron superconductivity in the strong coupling limit $\lambda \gg 1$ /3/.

For a polaron superconductor (PS)/4/:

$$T_c = 1.14 \, W \, (1-E_F^2/W^2)^{1/2} \, \exp\left(\frac{-2W}{V_0 + ZV_1 E_F^2/W^2}\right) \quad (6)$$

where V_0 and V_1 are on-site and inter-site polaron-polaron interaction by means of the local static lattice deformation of the order of $2g^2\omega$. For a bipolaron superconductor (BS)/3,6,9/

$$T_c = f(p)/m^{**} \quad (7)$$

where $f(p)$ is a function of concentration of the carries p, $f(p) - 3.3(p/2)^{2/3}$ for small p, and m^{**} is the effective bipolaron mass determined by

$$m/m^{**} = 2T_{nn} \Delta^{-1} \exp(-2g^2) \int_0^\infty dt \, \exp(-2g^2 \exp(\omega t/\Delta)-t), \quad (8)$$

where $T_{nn'} \sim D$ is an initial hopping integral, m is the band mass in a rigid lattics, and $\Delta = (2g^2\omega - V_c)$ is the bipolaron binding energy, V_c is a Coulomb repulsion between two polarons on the same or neighboring sites depending on what kind of bipolarons are formed: on-site or intersite local pairs.

As a result instead of a monotonous rise of Tc with the increase of the electron-phonon coupling, predicted by Eliashberg theory,/10/ the polaron theory of superconductivity predicts a rather narrow maximum T* in the dependence of Tc(λ) (Fig.1)

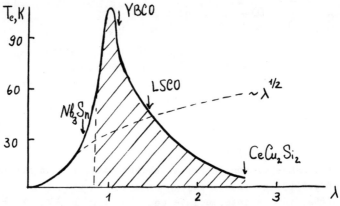

Fig.1. Dependence of the critical temperature on the electron-phonon coupling constant. The shading shows the area of polaron and bipolaron superconductivity. The dotted line corresponds to the Eliashberg theory.

T* may be estimated using Eq(6) and Eq(2) in the following way.
$$T^* \simeq 0.4g^2\omega \lesssim 0.5\omega \qquad (9)$$

In such a way one may achieve $T^* \simeq 500K$ by means of the interaction with a high-frequency oxygen vibration mode.

We can now proceed to the main properties of high-Tc lanthanium and yttrium cuprates, bearing in mind that BS is very reminiscient to superluidity of $He^{4/3/}$.

1. <u>High Tc</u>

Using Eq(7) with a reasonable values of $g^2 \simeq 2$; T_{mm}, Δ, $\omega \sim 0.1eV$ one can estimate
$$m^{**} \simeq (50 \div 100)\, m, \qquad (10)$$

which gives $T_c \simeq 100K$ for the appropriate hole concentration
$p = (0.5 \div 1) \cdot 10^{22} cm^{-3}$ (Hall measurements)

2. The London penetration depth λ_H

With the same values of m^{**} and p we obtain an enormously large

$$\lambda_H(o) = (\frac{m^{**}}{8\pi pe^2})^{1/2} \geqslant 3000 A^°, \qquad (11)$$

which agrees well with μsr-dates

The temperature dependence of $\lambda_H(T)$ is described by

$$\lambda_H(T) = \lambda_H(o) \ (1 + \frac{1}{3pm(2\pi)^3} \int d\bar{p} \ \bar{p}^2 \frac{\partial f}{\partial \mathcal{E}})^{-1/2}, \qquad (12)$$

where $f(\mathcal{E})$ a Bose distribution function. Using the exitation spectrum $\mathcal{E}(\bar{p})$ of BS /3/ one can easely obtain the power-law behaviour of $\lambda_H(T)$ for $T \ll T_c$:

$$\lambda_H(T) = \lambda_H(o) \ (1 + \kappa(T/T_c)^\alpha) \qquad (13)$$

with $3/2 < \alpha \leqslant 4$, $\kappa \sim 1$.
which seems to be the case for "1-2-3" /11/

3. Upward curvature of the upper critical field $Hc_2(T)$

Small bipolarons represent heavy interacting charged bosons.
The upper critical magnetic field of a charged Bose gas is given by

$$Hc_2(T) = H_{c1}(1 - t^{3/2})^{3/2}/t \qquad (14)$$

in the "clean" limit /12/, and by

$$Hc_2(T) = Hd \ (1 - t^{3/2})^{3/2}/t^{3/2} \qquad (15)$$

in the "dirty" limit /6/. Here

$$Hc = 0.18 \ /\phi_o m^{**} \ Tc \ \eta^{1/2},$$
$$Hd = 0.24 \ /\phi_o \ 1^{-1/2} (m^{**}Tc)^{3/2}$$

are the temperature independent constants determined by the free path of bosons, $\phi_o = \pi/e$, η is a gas parameter for weakly interacting bosons and l is the mean free path determined by the impurity-particle interaction.

Eq /14/ and Eq/15/ both predict a non-linear behaviour Hc_2 near T_c:

$$(Tc - T)^{1.5}$$

which well agrees with the experimental curve for the crystallic $YBa_2Cu_3O_{7-x}$ in a wide range of temperatures /13/.

4. Heat capacity

The first measurements /14/ of the Somerfeld constant $\gamma = C/T$ in the normal phase showed a rather small value

$$\gamma \lesssim 5 \text{ mJ/mol K}^2 \tag{16}$$

Now it is clear /15/ that it was an erroneous result due to the nonlinear $Hc_2(T)$ dependence Eq(15) and to a high value of resistivity of the first ceramic samples.

It turns out that new high-Tc superconductors have an enourmously high value of /15, 16/

$$\gamma > 30 \text{ mJ/mol K}^2 \tag{17}$$

which is of the same order of magnitude or even greater than γ of A-15 compands.

Taking into account that $\gamma \sim p^{1/3} m^{**}$, /6/ $\lambda_H \, p^{-1/2}(m^{**})^{1/2}$ and using Eq.(1) and Eq.(17) one can estimate

$$m^{**} > 20 \text{ m}, \qquad p \lesssim 10^{22} \text{ cm}^{-3} \tag{18}$$

which agrees satisfactorily with the estimation Eq.(10) and with the Hall measurements of p.

One of the most striking feature of LBSO, LSCO and YBCO crystals is the enormously high heat capacity jump./17/ Per one carrier in "1-2-3"

$$\Delta C/p \gtrsim 0.5 K_B, \tag{19}$$

where K_B is a Boltzmann constant.

Eq.(19) indicates that all the carries participate in the condensate formation as in liquid He^4, but not only their small fraction as assumed by the BSC-theory. Near Tc C(T) dependence should be λ-like as in liquid He^4, which seems to be the case for "1-2-3"/18/

5. The isotope effect

The first measurements gave

$$\alpha = - \frac{d \ln Tc}{d \ln M} \lesssim 0.2, \tag{20}$$

where M is the oxygen mass.
In the frame work of the BCS-approach a rather low value of α (Eq.20) could be explained by the anharmonicity of oxygen vibration modes as

well as Coulomb effects. In the polaron theory we have a practically zero α near the maximum of the dependence of $T_c(\lambda)$ (Fig.1).

On the other hand in the bipolaron limit ($\lambda \gg 1$) one can obtain an enormously high value of $\alpha > 0.5$ assuming that the high frequency oxygen vibration modes with $\omega \sim M^{-1/2}$ contribute mainly to g^2 (Eq.4). In this case using (Eq.7) with respect to (Eq.8) we obtain

$$\alpha = g^2 F(\Delta/\omega, 2g^2), \qquad (21)$$

where $F(x,y) = 1 + M^{-1}(1,1+x,y)(M(1,2+x,y) - x/y \frac{dM(1,1+x,y)}{dx})$

is a smooth function, which varies from $F(\infty,y) = 2$ to $F(0,y) = 1$, $M(a,c,z)$ is a confluent hyperbolic function. In such a way $g^2 \simeq 2$ gives $\alpha > 2$. So Eq.(21) explains, in general, the surprising experimental result of the Los-Alamos group[19].

6. XPS and EELS

XPS and EELS data permit us to propose the energy structure of high-Tc cuprates (Fig. 2) in which the O-2p energy band lies in the Hubbard gap $U \gtrsim 5eV$ between two Cu-3d Hubbard sub bands. This rather narrow O-2p ($D \lesssim 1eV$) band is responsible for the polaron and bipolaron formation.

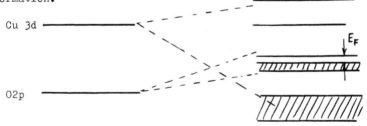

Fig.2 O-2p partially occupied narrow polaronic band between two Cu-3d Hubbard subbands.

Now we are in a position to answer the most crucial questions:
1. The high-Tc metallic oxides, as La-Ba(Sr)-Cu-O, Y-Ba-Cu-O, Bi-Sr-Ca-Cu-O, Tl-Ba-Ca-Cu-O as well as $Sr-TiO_3$, Ba-Pb-Bi-O, $LiTiO_2$ and others are bipolaronic by nature and may be described by the polaron-theory of superconductivity [3,4].
2. The measurable isotope-effect (20), tunnel spectroscopy, band-structure calculations and some other experimental data favour the determining role of the electron-phonon interaction in the high-Tc superconductors.

In conclusion it is worthwhile to note that the polaron theory explains the high-Tc superconductivity exclusively as a result of a favourable combination of bandwidth D and the electron-phonon coupling constant g^2 (Fig.1). So the layer structure as well as the presence of Cu ions seem to be not very important for the high Tc.

As an example, in the "heavy-fermion" systems ($CeCu_2Si_2$, UBe_{13}) the f-band is very narrow, so $m^{**} \gtrsim 1000m$ and according to Eq.(7) Tc is rather small ($< 10K$, Fig.1). It seems that La-Sr-Cu-O, Y-Ba-Cu-O, K-Bi-Ba-O have intermediate values of D and λ and are in the vicinity of the maximum of Tc(λ) (Fig.1).

The energy gap $\Delta(\xi)$ of the PS is defined by /4/

$$\Delta(\xi) = \frac{1}{2} \int_{-W-E_F}^{W-E_F} d\xi' \, N_p(\xi') \, V(\xi,\xi') \, \Delta(\xi') \, \text{th} \frac{(\xi'^2 + \Delta^2(\xi'))^{1/2}}{2T} \cdot (\xi'^2 + \Delta^2(\xi'))^{-1/2} \qquad (22)$$

where $V(\xi,\xi') = V_o + zV_1 \frac{(\xi + E_F)(\xi' + E_F)}{W^2}$, and $N_p(\xi')$ is the polaronic density of states.

For a smooth function $N_p(\xi)$ one may easely obtain using Eq.(22) $2\Delta/T_c \simeq 3.5$, which is close enough to the weak-coupling BCS value. On the other hand, in the BS ($\lambda \gg 1$) the energy gap is defined by the exitation spectrum of the charged interacting Bose-gas-bipolarons and may be considerably higher or lower than BCS value depending on the type of the partical-partical interaction:

$$0 \leq 2\Delta/Tc \lesssim 10 \qquad (23)$$

In such a way the ordinary electron-phonon interaction can produce high Tc as a result of the polaron narrowing of the band, which is not considered by the traditional theory of strong coupling superconductors, based on Migdal-Eliashberg equation. This interaction may be responsible for the other anomalous properties of high-Tc metallic oxides.

List of references

1. J.G.Beduorz, K.A.Müller, Z.Phys.B. **64**, 189(1986)
2. M.K.Wu, J.Asburn, C.J.Torng at al Phys.Rev.Lett., **58**, 908(1987)
3. A.Alexandrov, J.Ranninger, Phys.Rev.B., **23**, 1798 (1981), ibid. **B24**, 1164 (1981)
4. A.S.Alexandrov, Zh.Fiz.Khim. **57**, 273 (1983); Pis'ma Zh.Eksper. Teor.Fiz. **46**, addendum, 128 (1987)
5. P.W. Anderson, Science, **235**, 1196 (1987)
6. A.S.Alexandrov, J.Ranninger and S.Robaszkewicz, Phys.Rev. **B33**, 4526 (1986).
7. Yu.A.Firsov, Small-Radius Polarons. Transport Phenomena, in Polarony, Nauka, Moscow, 1985
8. A.S.Alexandrov, V.N.Grebenev and E.A.Mazur, Pisma Zh.Eksp.Teor. Fiz., **45**, 357 (1987).
9. A.S.Alexandrov and V.V.Kabanov, Fiz.Tverd.Tela (Leningrad), **28**, 1129 (1986).
10. G.M.Eliashberg, Zh.Eksp.Teor.Fiz. **37**, 366 (1960).
11. G.Aeppli et al, Phys.Rev. B **35**, 7129 (1987);
    C.T.Chu et al, Phys.Rev.B. **37**, 638 (1988);
    V.G.Grebennik et al, Pis'ma Zh.Eksp.Teor.Fiz., **46**, addendum,215 (1987)
12. A.S.Alexandrov, D.A.Samarchenko, S.V.Traven, Zh.Eksp.Teor.Fiz. **93**, 1007 (1987)
13. T.K.Worthington, W.J.Gaééagher, D.L.Kaiser, Holtzberg and T.R. Dinger, in Proceedings of the International Conference on high--temperature superconductors and materials and mechanisms of superconductivity, Interlaken, Switzerland, Feb.29-March 4 (1988), to be published.
14. D.K.Finnemore et al Phys.Rev. B**35**, 5319(1987)
15. M.B.Salamon and J.Bardeen, Phys.Rev.Lett, **59**, 2615 (1987)
16. M.E.Reeves, T.A.Friedman and D.M.Ginsberg, Phys.Rev. **B35**, 7207 (1987).
17. M.V.Nevitt, G.W.Grabbtree and T.E.Klippert, Phys.Rev. **B36**, 2398 (1987).
18. G.Deutscher in Proceeding of the International Conference on high-temperature superconductors and materials and mechanisms of superconductivity, Interlaken, Switzerland, Feb.29 - March 4, (1988), to be published
19. K.Otts et al, Superconductivity News 1, 7, (1988).

MOTT TRANSITION :

LOW-ENERGY EXCITATIONS AND SUPERCONDUCTIVITY

L. B. Ioffe

Landau Institute for Theoretical Physics, Moscow, USSR

A. I. Larkin

Landau Institute for Theoretical Physics, Moscow, USSR

Abstract

It is possible that a metal-dielectric transition does not result in changes of magnetic or crystallographic symmetry. In this case the Fermi spectrum does not change at the transition, but additional low-energy excitations appear which can be described as a gauge field with the same symmetry as electromagnetic one. In the case of a non half-filled band gapless scalar bose excitations also appear. Due to the presencs of additional gauge field the physical conductivity is determined by the **lowest** conductivity of the fermi or bose subsystems.

1. Introduction.

The recent discovery of high temperature superconductivity, and especially the observation that doping transforms the dielectrics into superconductors, has revived interest in the Mott-Habbard metal-dielectric transition. Qualitative properties of metal are well understood. Even for strong interaction the metal can be described by Fermi-liquid theory. The dielectric state is usually studied by a variational approach or in a mean field approximation. In the present problem both methods lack a small parameter; thus, it is hard (or even impossible) to estimate the accuracy and reliability of these methods.

The simplest model describing the system of strongly repulsing electrons is the Hubbard model:

$$H_H = \sum_{i,j} t_{ij} c^+_{i\alpha} c_{j\alpha} + \frac{1}{2} \sum_i U (c^+_{i\alpha} c_{i\alpha})^2 \qquad \alpha=1,2 \qquad (1)$$

If the electron band is half-filled and repulsion is strong: U>>t then the Hamiltonian (1) can be transformed to a simpler form:

$$H = \sum_{i,j} -J_{ij} c^+_{i\alpha} c_{j\alpha} c^+_{j\beta} c_{i\beta} \qquad J_{ij} = t^2_{ij}/U \qquad (2)$$

which is equivalent to the Hamiltonian of Heisenberg antiferromagnet:

$$H_A = \sum_{i,j} J_{ij} S_i S_j \qquad (3)$$

If the interaction J_{ij} is not zero only for nearest neihbours then the ground state of this Hamiltonian is a Neel antiferromagnet. The antiferromagnetic interaction of the next-nearest neigbours (frustration) results in the transition to the spin-liquid state, in which the mean spin at each site is exactly zero. Two questions immediately arise: can we prove the existence of such a state in a model with a small parameter and what is the spectrum of low energy excitations in this state ?

First we discuss the possibility of the existence of a spin liquid in the Heisenberg model (3) at large $S \gg 1$. At zero temperature and very large $S \to \infty$, thermal and quantum fluctuations are absent and the ground state is magnetic.

The increase of the next-nearest interaction results in the transition into a helical antiferromagnet. This transition can be a second order transition, in which case it can be described by Landau theory. At large but finite S the long-wave quantum fluctuations change the nature of the transition. The fluctuations are described by the effective action:

$$A\{n\}=\frac{1}{2}\int dt d^2x \{M\dot{n}^2 - \rho(\nabla n)^2 - \sigma_1(\Delta n)^2 - \sigma_2 \partial_x^2 n\, \partial_y^2 n\} \qquad (4)$$

where n is unit vector, pointing in the direction of the spin; $\Delta = \nabla^2$; M, ρ, σ_1, σ_2 are parameters which can be expressed through microscopic constants [1].

In the vicinity of the transition point $\rho \to 0$, the theory becomes logarithmic and the effective charge g that describes the spin-wave interaction obeys the renormalization group equations [1]:

$$dg/d(\ln R) = g^2 \qquad (5)$$

where R is the scale.

It is very important that the sign of the r.h.s. of the Eq.(5) is positive. This point differentiates the present problem from the phase transitions theory and makes it similar to the theory of two-dimensional (classic) magnets. The effective charge that was small at short distances ($g \propto 1/S$)

increases with scale. It means that the system has no magnetic long range order. Finite ρ cuts off the logarithmic divergences. Therefore, at $S \gg 1$ the spin liquid state exists in a narrow region $|\rho| \leq \exp(-1/S)$. Probably at $S=1/2$ this region becomes very large.

The renormalization group approach can not be extended to systems with strong interaction. Pomeranchuk [2] and Anderson [3] supposed that the excitations in the spin liquid are neutral fermions (spinons). At least two origins of these fermions are possible.

In the first scenario [4,5] the spinons are topological excitations of the soliton type which appear on the background of the short-range antiferromagnetic order. If in the effective action the term proportional to Hopf invariant (Chern-Simons term) is present with a half integer coefficient then the spin of these excitations is also half-integer. In this scenario the spin has a topological origin and does not interact with a physical magnetic field. The precise form of these excitations in the spin liquid is not clear, but in the Neel state they surely exist. However, in the Heisenberg model of antiferromagnet this coefficient is zero [1] in the ordinary Neel state so that skyrmions have an integer spin.

In the alternative scenario we shall show that Mott metal-dielectric transition has little effect on the spectrum of fermionic excitations, but the local gauge symmetry is restored. The creation operators of quasiparticles in the dielectric state are creation operators of electrons dressed by a phase factor. In the metal state the local gauge symmetry is broken, the phase factor acquires a mean value and fermionic Green function is not zero at large times and distances. In the dielectric state the gauge symmetry is restored, the mean value of the phase factor is zero ,the fermionic Green function decreases rapidly at large times and distances, auxilary gauge field screens out the electromagnetic one so that conductivity is zero. However, the two-particle correlation functions that correspond to a process without charge transfer (e.g. spin-spin correlation function) are the same as in metal. The collective excitations of the spinon liquid are described by a gauge field which is similar to electromagnetic and by a scalar bose field. In the case of a strong repulsion $U \gg t$ the number of holes in the whole system coincide with the number of bosons. We discuss the semiphenomenological theory of this state in the main body of this paper.

Specifically we find that the auxilary gauge field that appears in that problem can screen out the physical electromagnetic one so that the response to the physical electromagnetic field is determined by the largest of the resistance of fermi or bose subsistems. Thus the whole system is superconducting only if both fermi and bose subsystems are superconducting.

A quantitative description of this state can be obtained only in a model with a small parameter, for instance in a model introduced in [6,7]. This model is also described by Hamiltonian (1) (or by sum of Hamiltonians (1) and (2) with J_{ij} being regarded as an independent parameter), but α runs over N values: α=1..N (N>>1). The interaction of spinons with collective excitations is small in this model since it is proportional to 1/N so that quantitative results can be obtained. These results reproduce [6,11] the qualitative results of the semiphenomenological theory discussed in this paper.

2. Semiphenomenological theory and electromagnetic properties.

We start with a discussion of the form of the Hamiltonian that we choose to describe all important low-energy excitations of the system in the vicinity of the metal-dielectric transition and in dielectric state. Besides the real ("bare") electrons the new modes comprise the phase fluctuations of the auxiliary field Δ_{ij}. The physical meaning of this auxiliary field becomes transparent in the mean field approximation in which $\Delta_{ij} = J_{ij} \langle c_{i\alpha}^+ c_{j\alpha} \rangle$. The fluctuations of the amplitude of Δ_{ij} has a gap and, as we believe, has no impact on the qualitative properties of the low-energy excitations. In the model with large N (see the end of Introduction) these fluctuations are really small and can be neglected so that quantitative results can be obtained. The phase fluctuations of Δ_{ij} are always important, and here we shall discuss their effects.

To justify our form of semiphenomenological Hamiltonian of low-energy excitations we start from the bare electronic Hamiltonian for which we choose a slightly more general than a Hubbard form:

$$H = \sum_{i,j} t_{ij} c_{i\alpha}^+ c_{j\alpha} + \sum_{i,j} J_{ij} c_{i\alpha}^+ c_{j\alpha} c_{j\beta}^+ c_{i\beta} + \frac{1}{2} \sum_i U(c_{i\alpha}^+ c_{i\alpha})^2 \qquad (6)$$

in which besides Hubbard terms we include also additional superexchange interaction governed by independent parameter J_{ij} which has the same form as exchange interaction (2) describing the Hubbard model at $U/t \gg 1$ and a half-filled band. This model was proposed by Affleck and Martson [7].

To treat the superexchange interaction we introduce auxiliary field Δ_{ij} and employ Hubbard-Stratanovich transformation. We get the effective action:

$$S = -\int_0^\beta L d\tau$$

$$L = \sum_i \bar{c}_{i\alpha} \partial_\tau c_{i\alpha} + \sum_{i,j} \bar{c}_{i\alpha} c_{j\alpha} \exp(iAe_{ij})(t_{ij} - \Delta_{ij}^*) +$$
$$+ \sum_i [\tfrac{1}{2} U (\bar{c}_{i\alpha} c_{i\alpha})^2 - \mu \bar{c}_{i\alpha} c_{i\alpha}] \qquad (7)$$

$$\Delta_{ij} = \exp[i(\Phi_i - \Phi_j + a_{ij})]$$

where we exploit the path integral representation in the imaginary time τ, A is external electromagnetic field, μ - chemical potential. Below we shall not take into account the fluctuations of the amplitude $|\Delta_{ij}|$, thus we omitted the terms which depend only on $|\Delta_{ij}|^2$ in (7). The separation of the phase of Δ_{ij} into $(\Phi_i - \Phi_j)$ and a_{ij} is ambiguous, it should be determined from some auxiliary condition imposed on a_{ij} (choice of the gauge) which we discuss below. To treat the fermionic interaction remaining in (7) we introduce one more auxiliary scalar field ϕ_i:

$$L = \sum_i \bar{c}_{i\alpha} \partial_\tau c_{i\alpha} + \sum_{i,j} \bar{c}_{i\alpha} c_{j\alpha} \exp(iAe_{ij})(t_{ij} - \Delta_{ij}^*) +$$
$$+ \sum_i [(i\phi_i - \mu) \bar{c}_{i\alpha} c_{i\alpha} + \tfrac{1}{2} U^{-1} \phi_i^2] \qquad (8)$$

Then we perform the gauge transformation of variables:

$$c_i \to c_i \exp(+i\Psi_i)$$
$$\phi_i \to \phi_i - \dot{\Psi}_i \quad (9)$$

and get:

$$L = \sum_i \bar{c}_{i\alpha} \partial_\tau c_{i\alpha} + \sum_{i,j} \bar{c}_{i\alpha} c_{j\alpha} \exp(iAe_{ij}) \, [t_{ij}\exp(-i\Psi_i+i\Psi_j) +$$
$$+\exp(-ia_{ij})|\Delta_{ij}|] + \sum_i [(i\phi_i - \mu)\bar{c}_{i\alpha}c_{i\alpha} + \tfrac{1}{2}U^{-1}(\phi_i - \dot{\Psi}_i)^2] \quad (10)$$

The transformation properties of the fields a_{ij}, ϕ_i are equivalent to transformation properties of vector and scalar fields of lattice QED.

It is convenient to choose the gauge transformations (9) so that the resulting ϕ_i does not depend on time and replace the path integral over ϕ_i by path integral over $\dot{\Psi}_i(t)$ and ordinary integral over $\phi = \phi_o$. The constant ϕ_o should be determined from the condition of the free energy minimum:

$$i\langle c^+_{i\alpha} c_{i\alpha} \rangle = \tfrac{1}{U} \langle \dot{\Psi}_i - \phi_o \rangle \quad (11)$$

where $\langle .. \rangle$ means average with weight $\exp[-\int L d\tau]$ with L being defined by (10). The last term in the effective action (10) describes a system of non-interacting rotators governed by Lagrangian L_o:

$$L_o = \tfrac{1}{2U} \sum_i (\dot{\Psi}_i - i\phi_o)^2 \quad (12)$$

where we use real-time representation. The presence of the crossterm $\dot{\Psi}_i \phi_o$ in lagrangian L_o distinguishes these rotators from the ordinary ones. To obtain their energy spectrum we employ the Schrodinger representation. The canonical momentum $\mathfrak{M}$ conjugate to the variable Ψ_i is

$$\mathfrak{M} = \tfrac{1}{U}(\dot{\Psi}_i - i\phi_o) \quad (13)$$

Inserting the expression (13) for a canonical momentum

into Hamiltonian $H = \mathfrak{M}_i \dot{\Psi}_i - L$ we get $H = \frac{1}{2} U(\mathfrak{M} + i\phi_o/U)^2$. The wave function $Y(\Phi)$ should be periodic over Ψ so that eigenvalues of the operator $\mathfrak{M}$ are integers m and the corresponding energy levels are

$$\varepsilon_m = \frac{1}{2} U(m + i\phi_o/U)^2 \qquad (14)$$

In the rotator ground state $m = m_o$ where m_o is the integer which is closest to $-i\phi_o/U$ (it can be shown that ϕ_o obeying the condition (11) is purely imaginary so that $i\phi_o$ is real).

Now the equation (11) acquires a simple meaning: it ensures that the mean number of electrons in the system equals the mean value of operator $-\mathfrak{M}$.

The second term in the effective action (10) describes the interaction of rotators with each other and with fermions. For the qualitative analysis it is sufficient to replace the operator $c^+_{i\alpha} c_{j\alpha}$ in it by its mean $\langle c^+_{i\alpha} c_{j\alpha} \rangle$. The resulting term in the effective action describes interaction between neighbouring rotators.

We consider first the case of a strong repulsion $U \gg t$. In this case the interaction between rotators is small. For the half-filled band we can choose $i\phi_o/U = +1$. In this case the level spacing of each rotator is of the order of U, therefore in the ground state all rotators are in the same state $m = -1$. If holes are present then the mean value of $\mathfrak{M}$ is fractional, which implies that rotator wave function is a superposition of wave functions with $m = 0$ and with $m = -1$. Thus in this state we may choose $i\phi_o/U$ to be close to $+1/2$ so that the level spacing between levels $m = 0$ and $m = -1$ is of the order of t. The level spacing between other levels remains $U \gg t$. Therefore in this state each rotator can be described by two-level system. The excitations to the higher level are bose particles, they can be described by the operators S_i which are equal to the operator $\exp(-i\Psi_i)$ projected on the remaining two levels. If the density of holes is small then the number of bose excitations is also small and it is convenient to represent operators S_i, S_i^+ as a series over creation-anihilation operators of bose field. We make use of the Holstein-Primakoff transformation and get

$$S_i^+ = b_i^+ (1 - b_i^+ b_i) \qquad (15)$$

where we retain only the leading and the next term of the expansion over boson density. The projection of operator $\mathfrak{M}$ on the remaining two levels equals operator $1 - b_i^+ b_i$ describing the number of bosons. Thus in the bose representation the condition (11) acquires a simple form: $\langle c_{i\alpha}^+ c_{i\alpha} \rangle + \langle b_i^+ b_i \rangle = 1$ which is exactly the constraint inserted by hand in the slave boson approach [8]. It means that the number of bosons equals the number of physical holes. In the bose representation Hamiltonian of rotators becomes

$$H\{b\} = H_o\{b\} + H_{int}\{b\}$$
$$H_o\{b\} = \sum_{i,j} t_{ij} b_i^+ b_j + \varepsilon \sum_i b_i^+ b_i \qquad (16)$$
$$H_{int}\{b\} = -\sum_{i,j} t_{ij}^* [b_i^+ b_j^+ b_i b_i + b_j^+ b_j^+ b_i b_j] - \varepsilon \sum_i b_i^+ b_i^+ b_i b_i$$

where $t_{ij}^* = t_{ij} \langle c_{i\alpha}^+ c_{j\alpha} \rangle$, ε is the level spacing between m=0 and m=1 levels:

$$\varepsilon = \frac{1}{2} U + i \phi_o \qquad (17)$$

If the hole density is small then the gas approximation can be employed to study bose system (16). In this case only long-wave bosons are important, their bare spectrum follows from

Hamiltonian $H_o\{b\}$:

$$\varepsilon_o(k)=t^*(k)+\varepsilon=\varepsilon-t_o+t_1 k^2 \qquad (18)$$

At $\varepsilon > t_o$ there is no bosons in the ground state. At $\varepsilon < t_o$ the bare spectrum becomes unstable, but bose interaction described by $H_{int}\{b\}$ corresponds to repulsion of long-wave bosons, therefore in this case bose condensate is formed $<b> \neq 0$. In the gas approximation the density of this condensate is governed by the scattering amplitude Γ of two bosons:

$$n_c = |<b>|^2 = \frac{t_o - \varepsilon}{2\Gamma(\omega=0, k=0)} \qquad (19)$$

The scattering amplitude Γ at zero frequency and zero momentum can be obtained with logarithmic accuracy from a summation of ladder diagrams:

$$\Gamma = \frac{4\pi t_1}{\ln[t_o/(t_o-\varepsilon)]} \qquad (20)$$

In the leading order over $\{\ln[t_o/(t_o-\varepsilon)]\}^{-1}$ the total density of bosons n coincide with n_c. The difference between n and n_c is proportional to the next order of the small parameter $\{\ln[t_o/(t_o-\varepsilon)]\}^{-1}$:

$$n - n_c \simeq (t_o-\varepsilon)/4\pi t_1 \simeq n_c \{\ln[t_o/(t_o-\varepsilon)]\}^{-1} \qquad (21)$$

Thus in the present system the total number of bosons (and, thus the hole density) is zero at $n_c=0$ which means that at $n_c=0$ the band is half-filled. This is general result for the bose system with repulsion; whereas the attraction between bosons results in a macroscopic collapse. In paper [13] it was shown that for a system of one type bosons there is no alternative to one-particle condensate other than collapse. However an other scenario is possible if there are a few types of bose fields in the system. In this case a pair bose condensate can be formed:

$\langle b^{\alpha}b^{\beta}\rangle \neq 0, \langle b^{\alpha}\rangle = 0$. For instance, this is the case for a system consisting of a number of planes in which a weak tunneling between planes result in an effective attraction between bosons on adjacent planes, whereas bosons on the same plane still repel one another (this mechanism of boson attraction was proposed in [10]).

The appearance of these bose condensates results in the breakdown of the local gauge symmetry related with the fields a_{ij}, ϕ.

The averaging over bose and fermi fields leads to the effective action of electromagnetic field **A** and gauge field a_{ij}. The effective interaction of bose and fermi fields with gauge fields **A**, a_{ij} follows from (10):

$$L_{int} = \sum_{i,j} \bar{c}_{i\alpha} c_{j\alpha} \exp(i\mathbf{A} \mathbf{e}_{ij}) [t_{ij} b_i b_j^+ - \exp(-ia_{ij}) |\Delta_{ij}|] \quad (22)$$

The action (22) is invariant under gauge transformations $\mathbf{A} \to \nabla \theta_A + \mathbf{A}$, $c_i \to c_i \exp(i\theta_A)$ and $a_{ij} \to a_{ij} + \theta_{\alpha i} - \theta_{\alpha j}$, $c_i \to c_i \exp(-i\theta_{\alpha})$, $b_i \to b_i \exp(i\theta_{\alpha})$, which means that the charges of fermi field with respect to gauge fields **A** and a_{ij} are +1 and -1 correspondingly, the charges of bose field with respect to gauge fields **A** and a_{ij} are 0 and -1.

In the absense of holes the term proportional to t_{ij} in the effective action (22) can be omitted, so that the gauge fields are present in the action in the difference $\mathbf{Ae}_{ij} - a_{ij}$ only. In that case the effective long-range action of the gauge fields **A**, a_{ij} can be expressed through the fermionic polarization operator $\Pi_{\alpha\beta}$:

$$S\{A,a\} = \frac{T}{2} \int d^2k \sum_{\omega} \{[A_{\alpha}(\omega,k) - a_{\alpha}(\omega,k)]\Pi_{\alpha\beta}(\omega,k)[A_{\beta}(\omega,k) - a_{\beta}(\omega,k)]\} \quad (23)$$

Therefore in that case the average over a_{ij} results in the action which does not depend on the electromagnetic field at all. It means that in the absence of holes the electromagmnetic response is absent as it should in dielectric.

In the lowest order approximation over hole density the effective action of the gauge fields becomes:

$$S\{A,a\} = \frac{T}{2}\int d^2k \sum_\omega \{[A_\alpha(\omega,k) - a_\alpha(\omega,k)]\, \Pi_{\alpha\beta}(\omega,k)\, [A_\beta(\omega,k) - a_\beta(\omega,k)] + a_\alpha(\omega,k)\pi_{\alpha\beta}(\omega,k)a_\beta(\omega,k)\} \quad (24)$$

where the first term is generated by fermions that interact with both fields A, a_{ij} and the second is generated by bosons which interact only with field a_{ij}. It is convenient to single out from $\Pi_{\alpha\beta}$ and $\pi_{\alpha\beta}$ their longtitudinal and transverse parts:

$$\Pi_{\alpha\beta} = (\delta_{\alpha\beta} - \frac{k_\alpha k_\beta}{k^2})\Pi_1 + \frac{k_\alpha k_\beta}{k^2}\Pi_2$$

$$\pi_{\alpha\beta} = (\delta_{\alpha\beta} - \frac{k_\alpha k_\beta}{k^2})\pi_1 + \frac{k_\alpha k_\beta}{k^2}\pi_2 \quad (25)$$

The appearance of bose condensate means that π_1 remains non-zero at $\omega, k \to 0$: $\pi_1(\omega=0, k=0) = \rho_b$. At temperatures above the critical point of bose condensation $\pi_1(\omega \to 0, k=0) = \sigma_b \omega$. If the repulsion U is strong then the number of bosons equals the number of holes which implies, in particular, that σ_b becomes zero in the case of a half-filled band. At finite U we should take into account that the rotator should be described by two bose fields: one field corresponds to creation of excitations with $m > m_0$ and the other to excitations with $m < m_0$. The number of holes equals the difference between the numbers bosons of these two types. At finite temperature and half-filled band the conductivity becomes finite but exponentially small: $\sigma_b \propto$

exp(-U/2T) ,since the minimal energy of excitation energy in this state is U/2. If the band is half-filled the gap in bosonic spectrum which is nearly U/2 at U>>t decreases with U and becomes zero at $U=U_c \approx t$. At $U<U_c$ a bose condensate is formed which breaks the gauge symmetry.

Now we turn to fermionic excitations. The Green function $G_{ij}(t)$ of real electrons differ from the Green function of spinons by a phase factor:

$$G_{ij}(t)=<\bar{c}_{i\alpha}(t)c_{j\alpha}(0)\exp(i\Phi_i-i\Phi_j)> \qquad (26)$$

In a state with a one-particle bose-condesate the mean value of the phase factor is non zero at $t,r_{ij} \to \infty$ so that the residue of the pole of the electron Green function is also non-zero in this state, whereas in a state without bose condensate the mean value of the phase factor tends to zero at $t,r_{ij} \to \infty$. Thus in this state the residue of the electron Green function is zero on the Fermi surface. However gapless fermionic excitations (spinons) are present in this state

The exchange of virtual quanta of the field a_{ij} lead to the repulsion between fermions separated by a large distance. The exchange by virtual quanta of the bose field can lead to attraction at large distancesbut its strength is proportional to the density of bosons (i.e. density of holes) and is weak. Thus in the framework of the simplest one-plane model with equivalent sites the effective interaction of fermions is repulsive at large distances The exchange of short wave fluctuations of the Δ_{ij}-field usually result in short-range attraction between fermions.Thus in that system a weak superconductivity with a large correlation length is impossible, but the possibility of a strong superconductivity with a small correlation length ξ can not be excluded on such general ground and deserves a special study. If the system of fermions is normal then $\Pi_1(k=0)=\sigma_f\omega+O(\omega^2)$. In more complicated models the interaction between fermions can become attractive, in this case at low temperature the system of fermions becomes superconductive and $\Pi_1(k=0)=\rho_f+O(\omega)$.

For instance in the model which consists of many planes coupled by weak tunnelling the exchange of virtual bosons results in weak long-range attraction between fermions on adjacent planes. Since gauge field a_{ij} is purely two-dimensional the long-range repulsion between fermions on different planes is absent. Thus in this model the effective long-range interaction between fermions on adjacent planes is attractive and leads to superconductivity of the fermion subsystem. This superconductivity is unusual since it originates from off-diagonal pairing of fermions on adjacent planes. This form of pairing was originally proposed for layered materials in [14].

To get the effective action of the electromagnetic field which describes its interaction with the whole electronic system we should average the effective action (24) over long-wave fluctuations of the gauge field a_{ij}. Performing the averaging we get:

$$S\{A,\} = \frac{T}{2}\int d^2k \sum_\omega A_\alpha(\omega,k) P_{\alpha\beta}(\omega,k) A_\beta(\omega,k) \tag{27}$$

$$P_{\alpha\beta} = (\delta_{\alpha\beta} - \frac{k_\alpha k_\beta}{k^2}) P_1 + \frac{k_\alpha k_\beta}{k^2} P_2$$

where

$$P_i = \frac{\Pi_i}{\Pi_i} \frac{\pi_i}{+ \pi_i} \qquad (28)$$

The current j appearing as a reaction to the external electromagnetic field **A** is given by j=P **A** , thus formulae (26) means that the physical conductivity of the whole system is determined by the lowest conductivity of the fermi or bose subsystems.

This means that if bose subsystem is superconducting and fermi subsystem is not then the physical conductivity is finite and equals σ_f. In the opposite case if fermi subsystem is superconducting and bose subsystem is not, then the conductivity is also finite and equals σ_b. If both subsystems are superconductive then the superconductive density of the whole system is $\rho = \rho_f \rho_b / (\rho_f + \rho_b)$. If both subsystems have a finite conductivity then the resistivity of the whole system is sum of the resistivities of the subsystems: $\sigma^{-1} = \sigma_b^{-1} + \sigma_f^{-1}$. If the band is half-filled and $U > U_c$ then σ_b tends to zero at $T \to 0$ therefore the conductivity of the whole system is zero in this state at T=0 independent of the state of the fermion subsystem.

Now we discuss the quantization of the flux in the superconducting state $(\rho_f, \rho_b \neq 0)$. Generally, the free energy of the superconductive state can be expanded over variations of the phases of the order parameters of fermi and bose subsystems:

$$F = \frac{1}{2} |\nabla \psi_f - 2(\mathbf{A}-\mathbf{a})|^2 \rho_f + \frac{1}{2} |\nabla \psi_b - e^* \mathbf{a}|^2 \rho_b \qquad (29)$$

where $e^* = 1$ if bose condensate is one-particle condensate ($\langle b \rangle \neq 0$) and $e^* = 2$ if it is two-particle bose condensate ($\langle b^\alpha \rangle = 0, \langle b^\alpha b^\beta \rangle \neq 0$) (note that in our units the charge of electron is unity)

If $\rho_b > \rho_f$ then the vortices formed by fermi order parameter

ψ_f (so that $a=0$ and $\nabla\psi_b=0$) are more energetically favourable. In this case the charge which determines the flux quantization condition is 2, as usual. In the opposite case, if $\rho_b<\rho_f$ then the vortices formed by bose order parameter (so that $A=a$ and $\nabla\psi_f=0$) are more favourable. In this case the charge which determines the flux quantization condition is e^*.

3. Conclusions

Usually the metal-dielectric transition is accompanied by a change in crystallographic symmetry: e.g. lattice period doubling which occurs if an antiferromagnetic order parameter is formed or in the Pierls transition. In these transitions the fermionic spectrum is changed. In any case the metal-dielectric transition results in the restoration of the gauge symmetry in the dielectric state. In this paper we have discussed a semiphenomenological theory of the metal-dielectric transition which is not accompanied by changes in the fermionic spectrum, which is associated with the restoration of gauge symmetry in the dielectric state only. As a consequence of the transition gapless gauge excitations appear in the dielectric state which have the same transformation properties as the electromagnetic field.

In the framework of the semiphenomenological approach it is impossible to establish which state has the lowest energy. Even if the starting phenomenological Hamiltonian (7) is justified for the description of some material the possibility that a state with a period doubling is the genuine ground state is still open. This doubling can result from either a wave of the amplitude of Δ_{ij} (which result in molecular crystal state) or from oscillation of signs of Δ_{ij} such that the product of Δ_{ij} over the smallest lattice plaquette is -1 (this state can also be described as a state with a flux of a-field which is equal to π per each plaquette)[7,12]. The density of states on the Fermi-surface is zero (at the half-filled band) in both states so that the occurence of superconductivity in both states seems unlikely. This is the reason why we regard the uniform phase without the period doubling as the most interesting one.

It is possible that high-temperature superconductors based on copper oxides which easily become dielectrics at a small doping can be described by the semiphenomenological theory considered in this paper In this case a question arises: how is the superconductivity destroyed at T_c? Three cases are possible. If the transition temperature of the bose subsystem is above the transition temperature of the fermi one then the superconducting phase transition is driven by the transition of the fermi subsystem and its properties differ slightly from the properties of the usual BCS transition. If the transition temperature of bose subsystem is below the transition temperature of the fermi one then the superconductive phase transition is driven by the transition of the bose system, this scenario of the phase transition was proposed by Anderson and co-workers [3,8-10]. It is possible, finally, that phase transitions help each other and their critical temperatures are close.

References

1. L.B.Ioffe, A.I.Larkin Intern.Journ.Mod.Phys.B 2, 203 (1988).
2. I.Ya.Pomeranchuk ZhETP 11, 226 (1941).
3. P.W.Anderson Science 235, 1196 (1987).
4. F.Wilczek, A.Zee Phys.Rev.Lett. 57,2250 (1983).
5. I.E.Dzyaloshinskii, A.M.Polyakov. P.B.Wiegmann Phys.Lett.A 127, 112 (1988).
6. L.B.Ioffe, A.I.Larkin Phys.Rev.B to be published.
7. I.Affleck, J.B.Martson Phys.Rev.B 37, 3774 (1988).
8. G.Bascaran, P.W.Anderson Phys.Rev. B37, 580 (1988).
9. P.W.Anderson, G.Bascaran, Z.Zou, T.Hsu Phys Rev.Lett. 58, 2790 (1987).
10. J.M.Wheatly, T.C.Hsu, P.W.Anderson Phys.Rev. B37, 5897 (1988).
11. L.B.Ioffe, A.I.Larkin ZhETP to be published.
12. G.Kotliar, Phys.Rev.B 37,3664 (1988).
13. P.Noziers,D.Saint James, J Phys (Paris) 43, 1133, (1982).
14. K.B.Efetov, A.I.Larkin ZhETF, 68, 155, (1975).

ELECTRON-EXCITON INTERACTION AND LOCAL PAIRING IN HIGH-T_c SUPERCONDUCTORS

I.O. Kulik, A.G. Pedan

Institute for Low Temperature Physics and Engineering,
Ukrainian SSR Acad. Sci., 47 Lenin Ave., 310164 Kharkov,
U.S.S.R.

ABSTRACT. The model of superconductivity relating to the formation of pairs of holes at adjacent oxygen atoms, situated perpendicularly to the copper-oxygen chains in $YBa_2Cu_3O_{7-x}$ compound is considered. In slave boson representation for negative-U periodic Anderson model, the interaction of localized hole states with the conduction band is taken into account. The model is reduced to the local-pair scheme of high-T_c superconductivity with conduction band assisted transfer of localized pairs of holes.

1. Introduction

Theoretical interpretation of high-T_c superconductivity in oxides requires investigation of the electronic structure of metal-oxide compounds. Numerous spectroscopic works [1-4] evidence that these materials behave as systems with strong interelectron correlation, and that they cannot be described in terms of the simple one-particle (Fermi liquid) picture. Besides, the absence or very small magnitude of the isotope effect [5, 6] suggests the possibility of non-phonon mechanism of superconductivity. Analogous conclusions result from numerical calculations by Weber and Mattheiss [7]. Because of this, at present, main efforts are directed to the analysis of the models with electronic mechanisms of pairing.

The electronic states of copper-oxygen subsystem in high-T_c compounds are believed to be of main importance. It was supposed [8, 9] that superconductivity in the high-T_c compound $YBa_2Cu_3O_{7-x}$ is connected with the formation of local pairs of holes at oxygen ions in the "quasimolecules" $Cu(1)(O(4))_2$ in the ionization state $O^- - Cu^{2+} - O^-$ (see Fig. 1), which are situated perpendicularly to the copper-oxygen chains $Cu(1) - O(1) - Cu(1) - O(1) \ldots$ and to the copper-oxygen $O(2) - Cu(2) - O(3)$ planes.

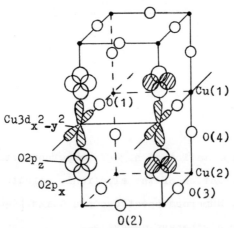

Fig. 1. Electronic configurations of copper and oxygen atoms in the $YBa_2Cu_3O_{7-x}$ lattice.

The hypothesis is supported by the compact disposition of O(4), Cu(1), O(4) atoms as according to [10] the distance C_1

= O(4) - Cu (1) = 1.826 Å is much smaller than the distance C_2 = Cu(2) - O(4) = 2.406 Å, and it is also slightly less than in-chain distance b/2 = Cu(1) - O(1) = 1.960 Å.

2. Charge fluctuation induced local pairing

In the ground state of $Cu(1)(O(4))_2$ molecule, copper ion contains one hole in the orbital configuration $3d_{x^2-y^2}$ (Cu^{2+} state) whereas oxygen ions have fully occupied 2p shells (O^{2-}). This is the consequence of the inequality $\varepsilon_p > \varepsilon_d$, i.e., the hole energy on oxygen greater than the hole energy on copper. Doping of oxide superconductor or increase of oxygen contamination compared to the stoichiometric composition results in the appearance of extra holes. These holes, due to strong Hubbard repulsion between two holes on copper orbital ($U_{dd} > \varepsilon$, $\varepsilon = \varepsilon_p - \varepsilon_d$), fill 2p states of oxygen. Comparison of different orbital energies shows that holes should be localized on p_z orbitals of O(4) ions. Two holes, due to their mutual repulsion, tend to localize at different oxygen ions. Exclusion of the inessential high-energy states (two holes on the same orbital, two holes on p_x and $d_{x^2-y^2}$ orbitals, etc.) results in the Hamiltonian describing the quasimolecule Cu(1) $(O(4))_2$ [9]:

$$H = H_x^o + H_z^o + H_{int},$$

$$H_x^o = \varepsilon_p \sum_{i\sigma} n_{i\sigma}^x + \varepsilon_d \sum_{\sigma} n_\sigma^d - t \sum_{i\sigma}(p_{ix\sigma}^+ d_\sigma + h.c.),$$

$$H_z^o = \varepsilon_p \sum_{i\sigma} n_{i\sigma}^z + U_c \sum_{\sigma\sigma'} n_{1\sigma}^z n_{2\sigma'}^z,$$

(1)

$$H_{int} = U_1 \sum_{i\sigma\sigma'} n_{i\sigma}^z n_{i\sigma'}^x + U_2 \sum_{i\neq j}\sum_{\sigma\sigma'} n_{i\sigma}^z n_{j\sigma'}^x + V \sum_i \sum_{\sigma\sigma'} n_{i\sigma}^z n_{\sigma'}^d.$$

Here, $p^+_{i\alpha\sigma}$ is the creation operator of hole with spin projection σ on p_α orbital of i-th oxygen ion (α, β = x, y, z; i, j = = 1, 2), d^+_σ - the creation operator of hole on $d_{x^2-y^2}$ copper orbital, $n^\alpha_{i\sigma} = p^+_{i\alpha\sigma} p_{i\alpha\sigma}$, $n^d_\sigma = d^+_\sigma d_\sigma$ • U_c, U_1, U_2, V - corresponding Coulomb interaction energies, t - hybridization matrix element between $d_{x^2-y^2}$ and p_x orbitals. Hybridization between p_y, p_z and $d_{x^2y^2}$ is weak and is neglected in Eq. (1).

The p-d hybridization results in the formation of bonding, antibonding and nonbonding states of CuO_2. Interaction of $2p_z$ holes with charge-transfer excitations corresponding to the transition between bonding and antibonding states is the main source of coupling between holes and results in the formation of local pair of holes.

Diagonalization of H^o_x by the transformation to the new operators a_σ, b_σ, c_σ:

$$p_{1(2)x\sigma} = \frac{1}{2}\sqrt{1 - \frac{\varepsilon}{\omega}} \; a_\sigma \pm \frac{1}{\sqrt{2}} b_\sigma - \frac{1}{2}\sqrt{1 + \frac{\varepsilon}{\omega}} \; c_\sigma \; , \quad (2)$$

$$d_\sigma = \sqrt{\frac{1}{2}(1 + \frac{\varepsilon}{\omega})} \; a_\sigma + \sqrt{\frac{1}{2}(1 - \frac{\varepsilon}{\omega})} \; c_\sigma$$

gives us

$$H^o_x = \frac{1}{2}\varepsilon \sum_\sigma n^b_\sigma + \frac{1}{2}\omega \sum_\sigma N_\sigma \; , \quad (3)$$

where $N_\sigma = n^c_\sigma - n^a_\sigma$ and $\omega = \sqrt{\varepsilon^2 + 8t^2}$.

Substitution of (2) into H_{int} results in the interaction Hamiltonian whose nondiagonal part can be represented in the form

$$-\frac{Ut}{\sqrt{2}\omega} \sum_i \sum_{\sigma\sigma'} n^z_{i\sigma} \; (c^+_\sigma \; a_{\sigma'} + a^+_{\sigma'} \; c_\sigma) , \quad (4)$$

where $\bar{U} = U_1 + U_2 - 2V$.

The structure of (4) is the same as in the electron-phonon interaction Hamiltonian. In our case Eq. (4) describes interaction of holes localized on p_z orbitals, with excitons whose creation operators are $c^+_6 \cdot a_6$. Final diagonalization of the Hamiltonian is achieved by canonical transformation

$$H = \hat{O}^+ H \hat{O}, \quad \hat{O} = \exp\left\{\sum_6 \hat{\xi} (c^+_6 \cdot a_6 - a^+_6 c_6)\right\}, \quad (5)$$

where

$$\hat{\xi} = -\frac{1}{2}\tan^{-1}\lambda, \quad \lambda = \frac{2\sqrt{2}\,\bar{U}t\sum_{i6} n^z_{i6}}{2\omega^2 + \bar{U}\varepsilon\sum_{i6} n^z_{i6}}. \quad (6)$$

After performing this transformation and averaging over the ground state of excitonic subsystem $a^+_6 |0\rangle$ we obtain Hamiltonian describing p_z holes with energies renormalized due to the electron-excitonic interaction. We are interested in the interaction between holes on p_z orbitals of different oxygen ions. This quantity, by its definition, is

$$U = E(0, 0) + E(1, 1) - 2E(0, 1), \quad (7)$$

where $E(0, 0)$, $E(1, 1)$ and $E(0, 1)$ are energies of $Cu^{2+}(O^{2-})_2$, $Cu^{2+}(O^-)_2$ and $Cu^{2+}O^{2-}O^-$ states, respectively.

Examination of the experimental data [1-4] shows that in oxide superconductors the following inequality between characteristic energies is satisfied: $\bar{U} \gg |t| \gg \varepsilon$, U_c, i.e., $U < 0$ which corresponds to the attraction between p_z holes. The attraction energy is of the order

$$U = \sqrt{2}\,|t| \sim 1 - 3 \text{ eV}. \quad (8)$$

Therefore, at temperatures $T < 10^3$ K practically all p_z

holes are bound into local hole pairs.

In the above approximation, local pairs with parallel ($\uparrow\uparrow$, $\downarrow\downarrow$) and antiparallel ($\uparrow\downarrow$) spins have equal energies. The degeneracy is removed by the inclusion of Cu $3d_{x^2-y^2}$ - O $2p_z$ hybridization

$$H' = t' \sum_{i\sigma} (p^+_{iz\sigma} d_\sigma + d^+_\sigma p_{iz\sigma}). \qquad (9)$$

Perturbation calculation based on the inequality $t' \ll t$ stabilizes triplet pairing, at sufficiently low temperature. Note, however, that this interaction is too weak to stabilize triplet pairs at temperatures $T \sim T_c$.

3. Slave boson representation and superconductivity in negative-U periodic Anderson model

Due to strong on-site repulsion of holes in copper, U_{dd}, real transition of holes between different copper ions are forbidden (Mott-Hubbard insulating state). Doping of oxides results in the appearance of free holes. It was postulated in [11, 12] that superconductivity in $YBa_2Cu_3O_{7-x}$ appears as a result of coherent transitions of hole pairs from their localization positions on quasimolecules $Cu(1)(O(4))_2$, to the conduction band states in the copper-oxygen planes $Cu(2)-O(2)-O(3)$, and vice versa. The analysis has shown, on the basis of phenomenological pair - conduction band interaction Hamiltonian, that this model has many properties in common with the experimental data for oxide superconductors. In particular, nonmonotonic dependence of critical temperature on hole concentration was obtained.

In what follows we consider microscopic derivation and extension of the Hamiltonian used in [11, 12]. We start with the periodic Anderson model with attractive on-site interaction of electrons ($-U < 0$). In the slave boson representation [13, 14], the Hamiltonian of the model takes the form [15]

$$H = H_0 + H_1 + H_2,$$

$$H_0 = E_1 \sum_{j\sigma} S_{j\sigma}^+ S_{j\sigma} + E_2 \sum_j d_j^+ d_j + \sum_{k\sigma} \mathcal{E}_k a_{k\sigma}^+ a_{k\sigma} - \hat{N},$$

$$H_1 = \frac{1}{\sqrt{N}} \sum_j \sum_{k\sigma} v_k a_{k\sigma}^+ e_j^+ S_{j\sigma} e^{ikR_j} + \text{h.c.}, \quad (10)$$

$$H_2 = \frac{1}{\sqrt{N}} \sum_j \sum_{k\sigma} \sigma v_k a_{k\sigma}^+ S_{j-\sigma}^+ d_j e^{ikR_j} + \text{h.c.},$$

where

$$N = \sum_{j\sigma} S_{j\sigma}^+ \cdot S_{j\sigma} + 2 \sum_j d_j^+ d_j + \sum_{k\sigma} a_{k\sigma}^+ a_{k\sigma}. \quad (11)$$

Here d_j^+, e_j^+ are bose-operators for creation of the site filled with local pair (energy E_2) and empty site, respectively, and S_j^+ are fermi-operators for creation of the one-electron state with spin projection σ and energy E_1. Hubbard term is introduced by the equality $E_2 = 2E_1 - U$, $a_{k\sigma}^+$ is the creation operator for conduction band electron with spin σ, momentum $\vec{k}$ and energy $\mathcal{E}_k$. All operators d_j, e_j, $S_{j\sigma}$, $a_{k\sigma}$ are canonical. They are connected with the local site operators $c_{j\sigma}$ by the relation [16]

$$c_{j\sigma} = e_j^+ S_{j\sigma} + \sigma S_{j-\sigma}^+ d_j, \quad \sigma = \pm 1. \quad (12)$$

Therefore, the hybridization term is represented as a sum of two components, $H_1 + H_2$, in Eq. (10). Those correspond to Feinman graphs shown in Fig. 2 with "slave" bosons d_j, e_j and "auxiliary" fermions $S_{j\sigma}$. In case of on-site attraction, Fermi level (μ)

is situated in the vicinity of local state level E_d ($E_d = E_2/2$, the energy of a pair converted to one electron) whereas single-particle local state has energy $E_d + U/2$ which is substantially higher than Fermi energy.

Fig. 2. Slave boson representation of hybridization between local and conduction band states.

Analytical calculations can be performed within the canonical scheme, Eqs. (10), whereas reduction to the physical state is achieved by projecting to the subspace of states obeying the local constraint [14]

$$Q_j = e_j^+ e_j + \sum_\sigma s_{j\sigma}^+ s_{j\sigma} + d_j^+ d_j = 1 . \qquad (13)$$

This condition means that only states corresponding to empty site ($e_j^+ |0\rangle$), site filled with single electron with spin up ($s_{j\uparrow}^+ |0\rangle$) or spin down ($s_{j\downarrow}^+ |0\rangle$), and to doubly occupied site filled with the pair ($d_j^+ |0\rangle$) are permitted.

Superconductivity is connected with the transitions between localized -pair and band states. Such terms in the Hamiltonian [11, 12] are readily obtained by canonical transformation eliminating linear in v_k contributions to Eq. (10). Second-order terms in the interaction Hamiltonian are shown in Fig. 3 in canonical ("slave boson") representation.

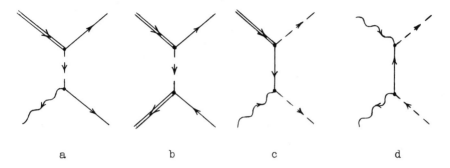

Fig. 3. Second order contributions to the pair - single electron scattering.

We write explicitly only contributions corresponding to Figs. 3a and 3c:

$$H^a_{int} = \frac{1}{N} \sum_j \sum_{kk'} g_{kk'} a^+_{k\uparrow} a^+_{-k\downarrow} e^+_j d_j e^{i(k-k')R_j} + h.c.,\qquad(14)$$

$$H^c_{int} = \sum_{ij\sigma} W_{ij} \, \sigma \, s_{i\sigma} s_{j-\sigma} e^+_i d^+_j + h.c.$$

Here in accordance with [8]

$$g_{kk'} = \frac{1}{2} v_k v_{-k'} \left\{ \frac{U}{(\varepsilon_k - E_d)^2 - U^2/4} + \frac{U}{(\varepsilon_{k'} - E_d)^2 - U^2/4} \right\} \qquad(15)$$

and

$$W_{ij} = \frac{1}{N} \sum_k \frac{v_k^2 (\varepsilon_k - E_d)}{(\varepsilon_k - E_d)^2 - U^2/4} e^{ik(R_i - R_j)} . \qquad(16)$$

In the local site representation, $e_j d^+_j$ (or $e^+_j d_j$) corresponds to the pair creation operator A^+_j of [11, 12]. Therefore, Fig. 3a and 4c correspond to the local pair transformation into two band states and into two localized single-particle states, respectively.

Hamiltonian $H_o + H_{int}^a$ for pairs and conduction band electrons is readily diagonalized by introducing quasiaverages

$$\Box = \frac{1}{N} \sum_k g_{kk}^* \langle a_{-k\downarrow} a_{k\uparrow} \rangle \quad , \quad \triangle_k = g_{kk} \langle e_j^+ d_j \rangle \quad (17)$$

vanishing at T_c. Critical temperature of superconducting transition equals

$$T_c = \frac{1-2v}{\ln\frac{1-v}{v}} \sum_k |g_{kk}|^2 \frac{\tanh\frac{\varepsilon_k - \mu}{2T_c}}{\varepsilon_k - \mu} \quad , \quad (18)$$

where v is average pair population per site determined by doping.

The term H_{int}^c describes transitions from excited non-paired local states to the localized paired states. It has naturally nothing to do with superconductivity as energy of two unpaired electrons and energy of local pair differ significantly. However, this term is essential in consideration of the relaxation from the excited state to local pair state which presumably is responsible for the narrow band cathodoluminescence observed in the yttrium ceramic at $T < T_c$ [17].

4. Conclusion

Formation of local pairs in the model of oxide metal considered is connected with the localization of holes at oxygen atoms in quasimolecules Cu (1) (O (4))$_2$. This is achieved by transition of oxygen atoms from the ionization state $(O^{2-})_2$ to the peroxide state $(O^-)_2$. Recently, peroxide species were detected in experiments on photoelectron emission spectroscopy in $La_{2-x}Sr_xCuO_4$ [18] and $YBa_2Cu_3O_{7-x}$ [19]. Moreover, the O^- ions show up in $YBa_2Cu_3O_{7-}$ in barium-oxygen planes [4], just in the planes at which O (4) atoms are situated. We think that peroxide-type local pairs are also present and determine physical properties of other metal oxide compounds. Note that peroxide species were identified in $BaPb_{1-x}Bi_xO_3$ and $SrPbO_3$ [20].

Slave boson technique proves to be useful in consideration of superconductivity in negative-U two-band Hubbard model (periodic Anderson model with on-site attraction). It permitted to studying, in the limit of weak coupling, of the nature of superconducting state which is described as the result of coexistence of Bose-Einstein condensation of local pairs and Cooper condensation of conduction band electrons, simultaneously appearing at T_c. Critical temperature of superconducting transition is proportional to the fourth power of the Anderson hybridization constant v_k and inversely proportional to the square of pair coupling energy, U (at $|v_k| \ll U$). We hope that same technique can permit consideration of superconducting transition in the limit of non-weak coupling which seems to be more realistic for actual superconducting oxide compounds.

Acknowledgments

One of the authors (I.O.K.) gratefully acknowledges Professor A.E. Ruckenstein's proposal of using slave boson technique for considering superconductivity in the local-pair theory, and for helpful discussions on the subject.

References

1 Bianconi A., Congiu Costellano A., De Santis M. et al. Z. Phys., 67, 307 (1987).

2 Horn S., Coi J., Shaheen S.A. et al. Phys. Rev. B, 36, 3895 (1987)

3 Tranquada J., Heald S., Moodenbaugh A et al. Phys. Rev. B, 35, 7187 (1987).

4 Steiner P., Hüfner S., Kinsinger V. et al. Z Phys. B, **69**, 449 (1988).

5 Batlogg B., Cava B.I., Jayaraman A. et al. Phys. Rev. Lett., **58**, 2333 (1987); Bourne M.F., Cromie M., Zettl A. et al. Ibid 2337.

6 Morris D., Kuroda R., Markelz A. et al. Phys. Rev. Lett. **61** (1988).

7 Weber W., Mattheiss L.F. Phys. Rev. B, **37**, 599 (1988).

8 Kulik I.O. Electronic transfer of local pairs and superconductivity in metal-oxide compounds. Proc. Adriatico Research Conf. "Towards theoretical understanding of high-T_c superconductors". Trieste, July 1988, World Scientific, Singapore, 1988.

9 Kulik I.O. and Pedan A.G. Sov. J. Low Temp. Phys. **14**, 700 (1988)

10 Jorgensen J.D., Beno M.A., Hinks D.G. et al. Phys. Rev. B, **36**, 3608 (1987).

11 Kulik I.O. Sov. J. Low Temp. Phys. **13**, 879 (1987).

12 Kulik I.O. Sov. J. Low Temp. Phys. **14**, 203 (1988).

13 Coleman P. Phys. Rep., **143**, 227 (1986)

14 Newns D.M., Read N. Adv. Phys., **36**, 799 (1987).

15 Newns D.M. Phys. Rev. B, **36**, 5595 (1987).

16 Kotliar G., Ruckenstein A.E. Phys. Rev. Lett., **57**, 1362 (1986).

17 Eremenko V.V., Fugol' I.Ya., Samovarov V.N. et al. JETP Lett. **47**, 533 (1988).

18 Sarma D.D., Rao C.N.R. J. Phys. C, **20**, L659 (1987).

19 Sarma D.D., Rao C.N.R. Sol. St. Comm., **65**, 47 (1988).

20 Ganguly P., Hegde M.S. Phys. Rev. B, **37**, 5107 (1988).

TIME DEPENDENCE OF MAGNETIZATION OF HIGH TEMPERATURE SUPERCONDUCTORS

V.B.Geshkenbein, A.I.Larkin

L.D.LANDAU INSTITUTE FOR THEORETICAL PHYSICS

ABSTRACT

Magnetization of high - T_c superconductors decreases logarithmically with time. There is a maximum in temperature dependence of the coefficient at this logarithm. By the assumption that there exist two kinds of pinning centers, this dependence is accounted in the Anderson theory for thermal creep of the Abrikosov vortices. The temperature dependence of the critical current is also discussed.

It was found that in new temperature superconductors magnetization decreases slowly logarithmically with time [1-5]. This decay follows from the Anderson model of flux creep [6].Another model proposed that the twinning boundaries are the Josephson weak links between the twins [7]. This model is like a spin glass model. In spin glasses the logarithmic time-dependence of magnetization is also observed. In experiments they usually find a ratio

$$r = -\frac{1}{M}\frac{dM}{d\ln t} \tag{1}$$

where M is magnetization of a superconductor. The temperature dependence of this ration has a characteristic maximum at 30-40 K in the field 500 G. In weak fields the position of this maximum shifts to higher temperatures. In spin glasses this maximum is not observed. The Anderson model also predicts a monotonous growth of r with temperature. So the authors of [3] claim that the decreasing of r with temperature doesn't definitely agree with flux creep model.

We show below that a small generalization of the Anderson model can explain this peak in the temperature dependence of r .This generalization consists in an assumption that there exist two kinds of pinning centers : many weak centers with a rather low activation energy and few strong centers with a high activation energy. At low temperatures weak centers give the main contribution into pinning and creep and the ratio rises monotonously with temperature. At high temperatures the vortex lattice near weak centers comes to thermal equilibrium during the experiment , the weak centers are excluded and the critical current is determined by strong centers where creep is much weaker.

If there is a current lower than the critical in superconductors then the vortex lattice is in a metastable state. Thermal fluctuations result in the fact that small areas of the lattice make thermal jumps.

If the current is close to the critical one then the barrier height is equal to

$$E = U_m \left(1 - \frac{J}{J_c}\right)^a \tag{2}$$

If we assume that the potential energy of each area which makes jumps can be approximately described by the one coordinate q then near the critical current this energy is a cubic parabola

$$U(q) = -\frac{U_m}{4} \frac{q}{q_m} [3(1 - \frac{J}{J_c}) - \frac{q^2}{q_m^2}] \qquad (3)$$

and the height of the barrier is determined by formula (2) with $a = 3/2$.

Probability of the thermal jump through such a barrier is proportional to $\exp(-E/T)$. As a result of these jumps the current in the sample decreases.

$$\frac{dJ}{dT} = \nu \exp(-\frac{E}{T}) \qquad (4)$$

where coefficient ν depends on the size of the sample. Solving this equation with a logarithmic accuracy we obtain

$$E = T \ln \omega t \qquad (5)$$

where $\omega = \nu \frac{dE}{dJ} / T$. Rough estimates give $\omega \sim 10^5 - 10^{10} (S^{-1})$.

After switching on a rather strong field there is a critical current flowing throughout the sample. This current decreases with time. From (2), (5) we obtain

$$\frac{J}{J_c} = 1 - (\frac{T \ln \omega t}{U_m})^{1/a} \qquad (6)$$

Magnetization of the sample is proportional to the current. Then we get:

$$r = -\frac{1}{M} \frac{dM}{d \ln t} = \frac{1}{a} (\frac{T}{U_m})^{1/a} (\ln \omega t)^{1/a - 1} \qquad (7)$$

Since the effective size of a potential well usually decreases with temperature increase, formula (7) gives a monotonous increasing of r. It holds if all pinning centers have an approximately equal value U_m. If there are two kinds of pinning centers with high barriers $U_m = U_1$, and low $U_m = U_2$, then (7) is valid only at a low temperature when $E = T \ln \omega t \ll U_2$. Near critical current formula (7) should be rewritten as

$$r = \frac{T^{1/a}}{a (\ln \omega t)^{1/a}} \frac{\sum_i \rho_i U_i^{1/a}}{1 - \sum_i \rho_i (\frac{T \ln \omega t}{U_i})^{1/a}} \qquad (8)$$

where ρ_i is a partial contribution of the i-th pinning centers to the current.

At temperature $T > U_2 / \ln \omega t$ weak pinning centers are excluded and we can use formula (7) again in which we must substitute U_1 for U_m. If $U_1 \gg U_m$, then ratio r at high temperatures will be smaller than at low temperatures where formula (8) works. There is a sharp jump of r at temperature $T = U_2 / \ln \omega t$ in this model. In the experiments there is a maximum and a smooth decreasing. There may be three reasons for this smooth decreasing. The first is that weak centers have dispersion U_i, at first the centers with the lower U_i. We can use formula (8) taking into account centers with $U_i > T \ln \omega t$ only. So when temperature increases there is an effective increasing of U_i and r drops. The second reason is that the magnetic field in the sample is inhomogeneous. If U_m depends upon a magnetic field then the pinning centers are gradually excluded. The distribution of the magnetic field depends on history. So r may depend on history too. The third reason is that when the current is much smaller than the critical one, then the height of the barrier isn't U_m but goes to infinity. It is so because the vortex lattice must be strongly reconstructed for passing the state with a lower energy than the initial one. There may be a high barrier between these states. For the field near H_{c1} where we can disregard the interaction between vortices, Vinokur and Feigel'man proposed that $E = U_m (J_c/J)^{1/4}$. It is based on the result [8]. Taking into account (5) we obtain

$$J = J_{c1} + J_{c2} (U_m / T \ln \omega t)^4$$

where J_{c1} is a contribution from strong centers which depends weakly on time. At high temperatures the second term is small and the ratio $r \propto T^{-4}$. There is a quantitative theory of creep today, especially for the case of a collective pinning, so it is difficult to say which reason gives the main contribution to a rather smooth decreasing of r with temperature increase. In any case the peak of r is near the temperature $T \sim U_m / \ln \omega t$. It is possible that with the rise of a magnetic field the effective U_m decreases and the maximum position shifts towards a low temperature region which corresponds to the experiment. The value of the ratio r in a maximum is equal to const $/\ln \omega t$. For characteristic lengths of the experiments $t \sim 1$ min $\omega t \sim 10^6 - 10^{10}$, $\ln \omega t \sim 10-20$ So, $r_{max} \sim 10^{-1}$, what agrees with the experimental value.

The physical reason for pinning of the vortices in oxygen superconductors may be randomly placed atoms of oxygen. One atom interacts weakly with a vortex lattice and cannot form a metastable state. But a high concentration of randomly placed atoms destroys the long range order in the vortex lattice [9]. The size of the area with a short range order depends on pinning force and elastic moduli of the lattice and is considerably larger than the distance between pinning centers. The number N of pinning centers in the area with a short range order is large. Such areas are weakly correlated and at thermal jumps they jump independently. So the height of the potential barrier at the motion of such area through the randomly placed pinning centers is proportional to $\sqrt{N}$ and may be rather large, although the critical current in such collective pinning is usually small [10,11].

Another reason for pinning may be the intersection of twinning boundaries [12]. In this case a critical current must depend on a number and structure of such boundaries. Such centers cause a plastic deformation of a vortex lattice and for them the Labusch criterion [13] is fulfilled. In this case there is a single-particle pinning when the mean force is proportional to the number of centers. Such centers may be large clusters of oxygen atoms in an area with the size ξ or some defects which can be formed, for instance, by irradiation of the sample by accelerated particles [14]. We don't know for which centers the pinning energy is high and for which it is low.

We consider above a picture of a thermal creep of the Abrikosov vortices. It is possible that the twinning boundaries are weak links [7]. In this case relaxation is determined by the motion of the Josephson vortices. A qualitative picture in this case is the same as in the motion of the Abrikosov vortices. But a quantative dynamics of the random Josephson medium has not been sufficiently studied. The thermodynamical properties of granular superconductors were studied in many works (for example, [15-17]. Dynamical properties were studied in [18]. But the authors of this paper considered the model of the Josephson junctions with a large radius of interaction and the temperature near T_c. Mathematically the model of granular superconductors in a magnetic field is like a spin glass model. But a direct comparison of the superconductor experiments with the experimental results on spin glasses and some theoretical results [19] is impossible, since a magnetic moment and magnetic field in these systems have a different physical meaning.

It is known [20, 21] that in high temperature superconductors the critical current obtained from magnetic measurements drops rapidly with temperature increase. For example, at $T = 45$ K $= T_c/2$ the current is by a factor of 10 smaller than at 4.2 K. Some authors write about the exponential deoendence of magnetization on temperature. Usually all parameters of superconductors approach constant values of $T \ll T_c$. So such a strong temperature dependence of the current in this region is not clear. Anderson [6] showed that a creep results in decreasing of magnetization with temperature. This is due to the fact that the current in the sample during the measurement ($\sim$ 1min) decreases due to the creep. For usual superconductors estimates show that this decreasing is negligibly small [12] and creep affects very weakly the temperature dependence of a current. For high superconductors we have other values of the parameters and the estimates must be done again. From (6), (7) we get

$$\frac{J - J_c}{J_c} \sim r(T) \ln \omega t$$

Usually, $\ln \omega t \sim 10 - 20$. For high temperature superconductors at $T \sim 10-20$ K, $r(T) \sim 0,05$, thus, from this simple approximation we see that at $T \sim 10-20$ K the difference between the measured current and the critical one is of the order of the critical current itself. Here and above by the critical current we imply the maximal current without fluctuations. This current determines magnetization at the initial moment (about $\sim 10^{-5}$ s). We can measure this current from current voltage curves. In such experiments the current is determined from the appearance of the threshold voltage V_c which is usually large what corresponds to a small time interval $\sim 10^{-5}$ s. Direct transport measurements of the current are less sensitive to creep than magnetic measurements. We don't know about resistive measurements of a critical current in single crystals of high temperature superconductors (the results of measurements on polycrystals are determined by weak links between granules). The current measured on films (from I–V curves) has the same order of magnitude as in single crystals and depends weakly on temperature at $T \ll T_c$. It agrees with hypothesis of a strong creep, but we don't know about any results of magnetic measurements on the films. The fact that the current decreases during experiments and can be much smaller than the critical one doesn't contradict the fact that the observed change of the current is rather small, because the time dependence of the current is logarithmic. The similar reasons for a strong creep in high T_c superconductors were discussed in [23].

At creep the activation energy $E = U_m(T) \cdot E_1(J/J_c(T))$. When $J \to J_c$, $E_1(J/J_c)$ is determined by (2). However, at small currents this dependence may be different. The problem for the theory is to determine current dependence of activation energy at any currents. As we discussed above, $E_1(J/J_c)$ may go to infinity when $J/J_c \to 0$. If we know the dependence of energy on current, then from (5) we can find temperature and time dependences of the measured current. For instance:

$$r = -\frac{T}{J\frac{dE}{dJ}} = -\frac{E(J)}{J\frac{dE}{dJ}\ln\omega t} \tag{9}$$

Assuming that at low temperatures ($T \ll T_c$) U_m and J_c are temperature-independent, the temperature dependence of the measured current is determined by the creep only. In this range we can connect time and temperature dependences of the measured current. Differentiating (5) over time and temperature we obtain

$$T\frac{dJ}{dT} = \ln\omega t \cdot \frac{dJ}{d\ln t} \tag{10}$$

In this formula all the derivatives are taken at the same time moment. We can disregard the temperature dependence of ω in logarithm because of the large value of this logarithm itself. For this reason it is not so important whether magnetization is measured in a minute or an hour and $\ln\omega t$ can be regarded as a constant. There is only one paper where we can find time and temperature dependences of the critical current for the same sample. [2]. These data agree with (10). Further investigations are necessary to estimate the role of the creep in high T_c superconductors.

The authors thank M. Feigel'man for helpful discussions.

REFERENCES

1. A.C.Mota, D.Visani, K.A.Muller, J.G.Bednorz Phys.Rev.B 36, 401 (1988).
2. Tuominen, A.M.Goldman, M.L.Mecartney Phys.Rev. B 37 548 (1988).
3. M.Tuominen, A.M.Goldman, M.L.Mecartney in materials of Interlaken Conference on High T_c superconductivity.
4. Sang B NAM In materials of Interlaken Conference on High T_c superconductivity.
5. Klimenko A.G., et al, pis'ma Zh.46, suppl., 196 (1987).
6. P.W.Anderson Phys.Rev.Lett. 9, 309 (1962).
7. G.Deutscher, K.A.Muller Phys.Rev.Lett. 59, 1745 (1987).
8. L.B.Ioffe, V.M.Vinokur J.Phys.C 20, 6149 (1987).
9. A.I.Larkin Zh.Exp.Teor.Fiz. 58, 1466 (1970).
10. A.I.Larkin, Yu.N.Ovchinnikov Zh.Exp.Teor.Fiz. 65, 1704 (1973).
11 A.I.Larkin, Yu.N.Ovchinnikov J.Low Temp.Phys. 34, 409 (1979).
12. V.B.Geshkenbein to be published in Zh.Exp.Teor.Fiz. 84, (9) (1988).
13. R.Labusch, Crystal Lattice Defects 1, 1 (1969).
14. J.R.Cost, J.O.Willis, J.D. Thompson, D.E.Peterson, Phys.Rev.B 37, 1563 (1988).
15. L.B.Ioffe, A.I.Larkin Zh.Teor.Fiz. 81 707 (1981).
16. J.Rosenblatt, P.Reyral, A.Raboutou.Phys.Lett. A, 98, 463 (1983).
17. C.Ebner, D.Stroud Phys.Rev. B31, 165 (1985).
18. V.M.Vinokur, L.B.Ioffe, A.I.Larkin, M.Feigel'man Zh.Exp.Teor.Fiz. 93, 343 (1987).

19. J.R.L. de Almeida , E.J.Thouless — J.Phys.A 11, 983 (1978).
20. G.W.Crabtree, J.Z. Liu, A.Umezawa et al Phys.Rev. B 36, 4021 (1907).
21. S.Senoussi , M.Oussena,? (Collin , I.A.Campbell Phys.Rev. B 37, 9762 (1988).
22. A.M.Campbell , J.E.Evetts Adv.Phys. 21 , 199 (1972).
23. A.M.Yeshurn , A.P. Malozemoff Phys.Rev.Lett . 60, 2202 (1988).

THE BERNOULLI EFFECT IN HIGH-T_c SUPERCONDUCTORS

A.N. Omel'yanchuk

Institute for Low Temperature Physics and Engineering, 310164 Kharkov, Lenin's ave. 47, USSR.

The nonuniform distribution of current density in superconductors gives the Bernoulli effect [1] - the appearance of electric potential difference (Bernoulli potential) between points with different values of the superfluid velocity $\vec{V}_s$. The Bernoulli potential compensates the current-induced change in the chemical potential of the electrons, so that in the thermodynamic equilibrium supercurrent state the full electrochemical potential remains constant in space. The Bernoulli effect is as universal as the Meissner effect, since owing to the latter the supercurrent in a bulk sample is always distributed nonuniformly. So, to describe the behaviour of equilibrium superconductors in external magnetic field (or external driving supercurrent) we generally have the coupled system of magnetostatic equation for distribution $\vec{V}_s(r)$, and electrostatic equation for electric field $\vec{E}(r)$ produced by the Bernoulli effect. The corresponding magnetoelectrostatic effects is usually very small (though measurable) because the dimensionless coupling constant is of the order of T_c/μ (T_c is the critical temperature, μ the chemical potential), but they can be appreciable in the case of high T_c (or small μ). In this paper, we investigate some specific features of the Bernoulli and related effects in high - T_c superconductors.

The Bernoulli effect in high - T_c superconductors can be described on the fenomenological level, irrespective of the microscopic model of superconducting state. We must supplement the equations of macroscopic electrodynamics with corresponding material relations for current and charge densities. For current - induced charge density we can write

$$\delta\rho(\vec{z}) = -\frac{1}{4\pi\lambda_D^2}\left[\Phi + \beta(\vec{V_s})\right]. \qquad (1)$$

Here λ_D is the screening length; Φ and $\vec{V_s}$ are the gradient invariant potentials: $\Phi = \varphi + \frac{\hbar}{2e}\frac{\partial \chi}{\partial t}$, $\vec{V_s} = \frac{\hbar}{2m^*}\left(\nabla\chi - \frac{2e}{c}\vec{A}\right)$. The function $\beta(\vec{V_s})$ describes the current-induced shift of the chemical potential of the electrons. The microscopic expression for $\beta(\vec{V_s})$, based on the BCS theory, was obtained in[2] (see also [3]). In fenomenological theory $\beta(\vec{V_s})$ is some function, in which we can separate two parts. One part is associated with the kinetic inductance of the current state, and the other is due to current - suppression of the energy gap Δ, $\beta = \beta_\alpha + \beta_\Delta$:

$$\beta_\alpha = \frac{\alpha j^2(\vec{z})}{2e}, \quad \beta_\Delta = -\frac{1}{e}\frac{\partial \mu}{\partial \Delta}\delta\Delta(\vec{V_s}), \qquad (2)$$

where $\delta\Delta(\vec{V_s}) = \Delta(\vec{V_s}) - \Delta(\vec{V_s}=0)$ is the variation of the order parameter with the current.

In the case of $T \ll T_c$

$$\beta_\alpha = \frac{m^* V_s^2}{2e}, \quad \beta_\Delta = 0. \qquad (3)$$

Let us study the manifestation of the Bernoulli effect in two situations. The first one is the current-controlled contact potential difference between the superconductors [3], and the second one is the electrical polarization of a superconductor placed in a stationary magnetic field [4].

In the approximation of local electrical neutrality we put $\delta\rho(\vec{r}) = 0$ and obtain the expression for the Bernoulli potential $\Phi(\vec{r}) = -\beta(\vec{V_s}(\vec{r}))$. Let us consider the geometry depicted in the Figure. The bulk superconductor S_1 is in contact

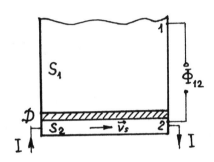

with the thin superconducting film S_2 through the dielectric layer D. The superconducting current I passes through the film S_2, leading to a shift of the chemical potential in S_2 and the appearance of a contact potential difference Φ_{12} which is the Bernoulli potential $\Phi = -\beta(\vec{V_s})$ between point in S_1 (where $\vec{V_s} = 0$) far from the region of the contact and the film S_2 (with a current homogeneously distributed over the film thickness and a corresponding value of $\vec{V_s}$). It is important to stress, that the Bernoulli potential and the corresponding contact potential difference Φ_{12} cannot be measured with a voltmeter (see discussion in [5,6]).

It was measured by the Kelvin capacitance method in the experiments of Ref.6. In our situation shown in the Figure, the potential difference Φ_{12} can be measured directly with a voltmeter in the quasistationary regime, in which the current I in the system varies with the frequency ω, which is small as compared to $\Delta/\hbar$ and τ_ε^{-1} (τ_ε is the inelastic relaxation time) and is large in comparison with the frequency $\omega_0 = 1/RC$ (where R is the resistance of the contact and C is its capacitance). Such experiment was carried out on an ordinary superconductor (Indium)[7], and yielded $\Phi_{12} \lesssim 10^{-8} V$. It would be interesting to conduct similar experiment on high - T_c metal-oxide compounds. The trivial reason is the increase in the effect with rising T_c, thus for $T_c \sim$ 100 K we have in the BCS model $\Phi_{12} \lesssim T_c \frac{T_c}{\mu} \sim 10^{-4} V$. Then, in the case of narrow band superconductors, particularly in the local - pair model[8], the shift of $\delta\mu$ does not contain the small factor T/μ, and we can expect $\Phi_{12} \sim T_c \sim 10^{-2} V$. So, the measurement of the Bernoulli potential is a probe for the nature of the superconducting state in metal-oxide compounds.

Now we shall proceed to magnetoelectrostatic effects, occuring when the equations of the macroscopic electrodynamics involves material relation (1). For distribution of electrostatic potential Φ in superconducting body we have the Poisson equation

$$\nabla^2 \Phi - \frac{1}{\lambda_D^2} \Phi = \frac{1}{\lambda_D^2} \beta(\vec{V_s}), \qquad (4)$$

which must be solved jointly with the equation for distribution of $\vec{V_s}(\vec{r})$ (the London equation in the simplest case).

The external magnetic field $\vec{H}$ induces Meissner currents in the superconductor, and hence, according to (4), the electric fields $\vec{E} = -\nabla\Phi$. So, taking into account the Bernoulli effect, a superconductor placed in a stationary magnetic field becomes electrically polarized. Such polarization was investigated in Ref 4. The electric polarization of surface layer with thickness λ_L (λ_L is the London penetration depth) is of the order of $P \sim \frac{1}{ne\lambda_L}(\frac{H^2}{8\pi})$ (n is the concentration of electrons). The picture is analogous to piezoelectric effect, the role of mechanical pressure being played by the magnetic pressure $\frac{H^2}{8\pi} \sim 10^6$ dyne/cm^2 (for $H \sim 10^3$ G) and the corresponding "piezoelectric" modulus is equal to $d_s = \frac{1}{ne\lambda_L} \sim 10^{-8}$ cm/dyne$^{1/2}$.

In conclusion, the discussed here current controlled contact potential difference and the magnetoelectrostatic effects caused by the Bernoulli potential are pronaunced in superconductors with high critical parameters (T_c, H_c), and can be used to investigate superconductivity in metal-oxide compounds.

References

1. F.London, Superfluids (John Wiley & Sons, Inc., New York, 1950), Vol.1, pp.55, §8.
2. K.M.Hong, Phys. Rev. B $\underline{12}$, 1766(1975).
3. A.N.Omel'yanchuk, S.I. Beloborod'ko, Sov.J.Low Temp. Phys $\underline{9}$, 572 (1983).
4. Yu.M. Ivanchenko, A.N.Omel'yanchuk, Sov.J.Low Temp.Phys. $\underline{11}$, 490 (1985).
5. T.K.Hunt, Phys.Lett. $\underline{22}$, N 1, 42 (1966).
6. J.Bok, J.Klein, Phys.Rev.Lett, $\underline{20}$, N 13, 660 (1968).
7. Yu.N.Chiang, O.G.Shevchenko Fiz. Nizk. Temp., $\underline{12}$, N 8, 816(1986).
8. I.O.Kulik, Sov.J.Low Temp. Phys. $\underline{14}$, N 2 (1987).

PREPARATION AND STUDY OF Tl BASED SUPERCONDUCTORS

A.I.Akimov[1], L.Z.Avdeev[2], S.V.Bogachev[3], B.B.Boiko[1],
M.M.Gaidukov[3], V.I.Gatalskaya[1], S.E.Demyanov[1], A.L.Karpei[1],
A.M.Klushin, L.A.Kurochkin[1], Yu.N.Leonovich[1], L.P.Poluchankina[1]
O.V.Snigirev[2], E.K.Stribuk[1], I.M.Starchenko[1]

1-Institute of Solid State and Semiconductor Physics
the BSSR Academy of Sciences, 2-Moscow State University,
3-Leningrad Electrotechnical Institute,

High-T_c TlBaCaCuO ceramics with T_c^{off}= 125.3 K have been prepared and our measurement data on X-ray, resistivity and magnetic characteristics, Hall effect, weak-field absorption, Josephson effect in thallium ceramics are reported.

Since the discovery /1/ of a new class of high-T_c superconducting bismuth ceramics with T_c= 22 K which contains no rare earth elements an intensive search for new superconductors has been going on. In /2-5/ the investigations on TlBaCuO and TlBaCaCuO samples obtained with the exception of /5/ from the reactions of melt-solid phase and annealing in oxygen atmosphere have been carried out. Two superconducting phases $Tl_2Ba_2CaCu_2O_8$ (2212) and $Tl_2Ba_2Ca_2Cu_3O_x$ (2223) with T_c^{on}= 100 K and T_c^{on}= 125 K have been isolated from samples by the authors /3,6,7/.

We prepared /8/ thallium based ceramics by solid phase synthesis from the mixture of oxides taken in stoichiometric ratio according to the formula $Tl_{1.4}BaCaCu_{1.5}O_y$. Part of the samples was obtained by sintering at T_s=(810-870)°C for 1-10hrs the other one - by a two-stage synthesis: after sintering the samples were crashed, pressed into pellets, sintered again at 840-880°C in aluminium crucibles. From the X-ray structure analysis data depending on temperature and time of annealing one obtains samples with pseudotetragonal structure, in this case both the samples with single phase and those with three superconducting phases. The phases have practically equal parameters $a = b$ = 3.85 Å (5.44) and C parameters are 29,30;31.70 and 36.50 Å, which is connected with different quantity of perovskite layers between Tl_2O_2.

The highest superconducting transition temperature of T_c^{off} = 125.3 K was achieved by the samples with the maximum parameter value. The measurement of temperature dependence of the lattice parameters of the phase with the parameter c = = 35.60 Å revealed an increase in C and a parameters, which corresponded to the V-like dependence peculiarity on the curve of temperature dependence of the thermal expansion coefficient $\alpha(T)$ (Fig.1) in the temperature range close to T_c viz. ~130-140 K. The anisotropic character of thermal expansion along the axis should be noted. The measurements were taken on the reflections (0016) and (200).

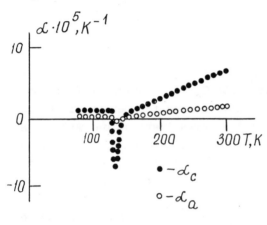

Fig.1

The character of the transition to the superconducting state and the volume of the superconducting phase depend essentially on the preparation conditions of the ceramics (Figs.2,3). Table gives the values of temperatures and annealing times of the samples obtained by one stage and two stage (with intermediate grinding) annealings.

The highest critical temperatures T_c^{off} were revealed for samples obtained by two stage annealing: sample N6-125.3 K, N7-120K, N8-115K, N9-115.6 K, N10-110.2K and the onset superconducting transition temperatures T_c^{on} for these samples were N6-133K, N7-124.5K, N8-123.5K, N9-120.5K and N10-118.5K, respectively. The largest volume of the superconducting phase is attained with one stage technology. It should be noted that the near surface layer is thallium poor and the thickness of this layer, depending on the density of the samples prior to annealing, varies from 0.3 to 0.8 mm.

Table

	One stage annealing		Two stage annealing	
1	870°C- 20 min	6	850°C-10 hrs	880°C -1 min
	860°C- 5 hrs			870°C -20 min
				860°C -5 hrs
2	860°C- 5 hrs	7	850°C-10 hrs	850°C -5 hrs
3	850°C- 1 hr	8	810°C-10 hrs	870°C -20 min
				860°C -5 hrs
4	850°C- 10 hrs	9	810°C-10 hrs	850°C -5 hrs
5	810°C- 10 hrs	10	810°C-10 hrs	840°C -5 hrs

Figs 2,3

We have measured the magnetic characteristics of Tl ceramics with a SQUID magnetometer between 10 and 200 K (to 100 Oe). Fig. 4 a shows the temperature dependences of magnetization for sample N7 obtained on warming in different magnetic fields, at $T < T_c$ the screening of the magnetic field is practically complete for fields $H < 10$ Oe. The temperature of disappearance of the diamagnetic response on warming in this case is

(124.8 + 0.5 K).

The curve of screening in the field of 0.9 Oe (Fig.4b) displays the peculiarity in the range of 110 K which is associated with the presence of a small quantity of a second semiconducting phase /7/. The Meissner effect value measured at H = 50 Oe amounted to 17 % of the screening in this field, which is characteristic of oxide ceramic superconductors. As follows from the magnetization measurements of Tl ceramics above T_c^{off} in the magnetic field of 100 Oe (Fig.5) the change of the magnetic moment with temperature has a Curie-Weiss character: $M = 1/(T-\theta)$, where $\theta = -60$ K. The calculation of the ef-

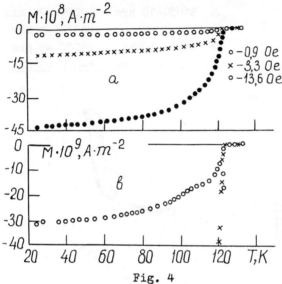

Fig. 4

fective moment yields the value $\mu_{eff} = 3\mu_B$ per formula unit. Also the weak-field microwave absorption was studied with a EPR SE/X-2543 "Radiopan" spectrometer in X-band on samples obtained by the technique as used for sample N7. The curve of this temperature dependence (Fig. 6) (the modulation amplitude of constant magnetic field of 2.5 Oe see inset

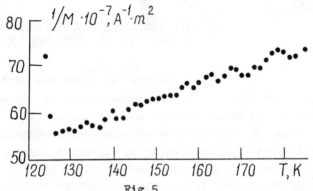

Fig.5

is characterized by (i) a sharp rise of the amplitude of the A signal when the temperature approaches T_c and (ii) by the constant A value when T decreases below T_c with subsequent broadening of the weak-field signal and the reduction of its amplitude. The weak-field microwave absorption at the frequency of 9.15 GHz in these samples can be explained by the granular nature of these ceramics consisting of clusters of weakly bound superconducting grains. The scatter of temperatures and widths of superconducting transition determined at the level of 0.5 A_{max} gives evidence of the presence of some phases, which is consistent with the results of magnetic measurements (Figs. 3,4).

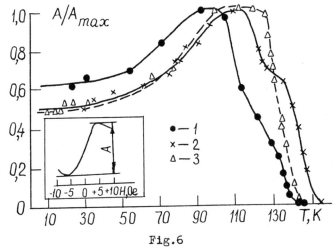

Fig.6

The measurements of the trapped magnetic flux and HF-SHF surface resistivity of Tl specimens are closely connected with the above mentioned studies of the microwave absorption in these ceramics. The samples N11 and N12 have the completion temperature T_c^{off} = 120 K and the resistivity of $\rho_{300\,K}$ = 0.6 mΩcm and 2mΩ cm, respectively. The results of the induction measurements of these samples (Fig. 7) revealing the presence of one more superconducting phase with T_c below 120 K correlate with the temperature dependences of the trapped flux B_{tr} at T T_c in the constant magnetic field B_0 = 11 Oe. The value of the frozen flux depends on the temperature of magnetic field switching off and the quality of the ceramics. For instance, 90% of B_0 freezes already at 80 K for sample N 11 and for sample N12 only 50% of B_0 does at the same temperature. In this case for sample N11 the value of the trapped flux remains constant with decrease of temperature of B_o switching off starting from ~80 K, while for sample N 12 - from 40 K. For comparison the temperature dependence of B_{tr}/B_o for YBACuO ceramics with ρ_{300K}= 3 mΩ cm is given in the same figure. The value of the crystal resistivity R_s was measured on Tl -samples at 37 MHz and 60 GHz. The essential difference between the R_s values for the normal and superconducting state in HF-range should be

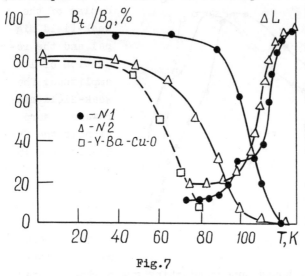

Fig.7

pointed out (in SHF-range this difference is insignificant). The investigation of the influence of the trapped magnetic flux on the surface resistance of Tl samples at various densities of magnetic flux and amplitudes $\tilde{H}_\perp$ of high frequency magnetic field of HF and SHF -ranges gives evidence of the effective freezing of the magnetic flux in the sample. The magnetic field was switched off at 30 K, which has ensured the maximum trapping of magnetic flux (Fig8) and the absence of

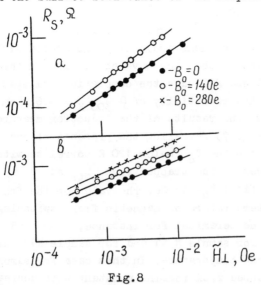

Fig.8

its freezing in the walls of the measuring resonator (T=4.2K). At low levels of $\tilde{H}\perp$ the values of dR_s/dB_o at frequency of 37 MHz are $2 \cdot 10^{-6}$ (N11) and $1 \cdot 10^{-5}$ (N12) Ω/Oe. The influence of high frequency field on R_s is more noticeable: $dR_s/d\tilde{H}\perp$ achieve $5 \cdot 10^{-2}$ and 10^{-1} /Oe (N11) and $6.1 \cdot 10^{-1}$ and $1.1 \cdot 10 \Omega$/Oe (12) for $B_o = 0$ and 14 Oe, respectively. Such dependences can be connected with the contribution of non-superconducting phases in Tl-ceramics.

The Hall effect measurements were made by a standard six-probe method to a precision of 10^{-8}V. Fig.9 gives the values of Hall voltage U_x which is positive at all temperatures above T_c and linear with both the magnetic field (up to 72 kOe) and the current (up to 10 A/cm^2). From our results it can be seen that the TlBaCaCuO ceramics conduct via holes in the normal state with carrier concentration of $(1.20 + 0.05) 10^{21}$ cm^{-3} at room temperature. Three large anomalies of U_x are observed between 70-130 K. These anomalies are similar to those found in LaSrCuO /9/ and YBaCuO /10/ systems. Near T_c in the range 120-130 K the grains contribution to Hall effect reduces rapid-

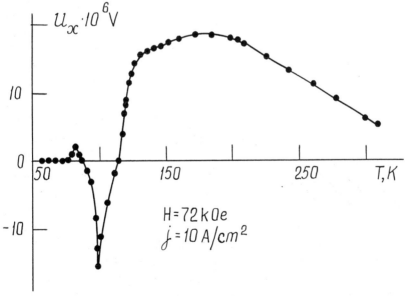

Fig.9

ly and becomes zero. At these temperatures the grain boundaries remain in the normal state. With further decrease of temperature the grain boundaries become superconducting ones. The anomalies between 78-100 K appear to be connected with two kinds of carriers at the grain boundaries: near 110 K the electron conductivity dominates in Tl-ceramics and at $T \sim 80$ K the hole conductivity does.

Of special interest are the observations of the Josephson effects on the TlBaCaCuO-TlBaCaCuO junctions at temperatures above 100 K. The Josephson junctions have been made by sawing up of thallium bars ($7 \times 1.5 \times 1.5$ mm^3 in size) until the critical current of 1-3 mA is acieved in liquid nitrogen. From the resistivity measurements (with an accuracy of $R = 10^{-4} \Omega$) $T_c^{off} = 118$ K has been obtained for the investigated ceramics /11/. The current voltage (I-V) characteristics have been measured with a four-probe method between 78 and 102 K. Fig. 10 represents a set of I-V characteristics at different microwave power levels P= 0 - 11 mW with F = 9.46 GHz at T= 98 K.

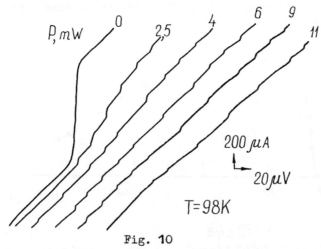

Fig. 10

The investigations of the temperature dependence of critical current I_c of the samples revealed that in the range from 78 to 102 K $I_{c*}(T) = 0.087 \times (T_{*c} - T)$ mA, where $T_{*c} = 105$ K.

At liquid nitrogen temperature the critical frequency of the junction $f_c = 73$ GHz, the normal resistance of the junction is $R_n = 0.075 \Omega$; at T = 98 $f_c = 41$ GHz, $R_n = 0.135 \Omega$. The increase of R_n with temperature can naturally be connected with the changes of the geometry of the superconducting range on warming which could occur due to

T_c nonuniformity of the cross section of the sample. The latter could account for the fact that T_c^* in the temperature dependence of T_c (T) of the junction is less than T_c of the initial sample.

The synthesis and comprehensive investigations of five-component superconducting TlBaCaCuO system have been carried out. Three phases with $T_c^{on} \approx 110$, 120 and 133 K have been found and a correlation of T_c with the crystal lattice parameters has been established.

The analysis of the magnetic, Hall effect, microwave absorption measurements gives evidence of high superconducting critical grains parameters and weak superconductivity of the material on the whole. This is obviously due to low critical parameters of the grain boundaries coupling the grains as the Josephson junctions.

The maximum value of $T_c^{off} = 125.3$ K achieved, the deviation from the linear trend of ρ (T) dependence as well as the anomalies in microwave absorption between (130-160)K speak for great potentiality of thallium based ceramics for future investigation.

References

1. Michel C., Hervieu M., Borel M.M. et al.//Z.Phys.B.1987. V.68.P.421.
2. Zheng L.L., Hermann A.M., El Ali A. et al.//Phys.Rev.Letts. 1988.V._0.P.937.
3. Hazen R.M., Finger L.W., Angel R.I. et al.//Phys.Rev.Letts. 1988.V.60.P.1657.
4. Gao L., Huang L.I., Meng R.L. et al.//Nature,1988.V.332. P.623.
5. Gingly D.S., Venturini E.L., Kwak I.F. et al.//Physica C. 1988.V.152.P.217.
6. Torardi C.C., Subramanian M.A., Calabrese J.C. et al.//Science, 1988.V.240.P.631.
7. Parkin S.S.P., Lee V.I., Nazzal A.I. et al.Preprint 1988.
8. Boiko B.B.,Akimov A.I., Gatalskaya V.I.,et al. Pis'ma v ZhETF, V.48.P.103.

9. Hindley M.F., Zettl A., Stacy A. et al.//Phys.Rev.1987. V.35 B. P.8800.
10. Yong Zh., Qirul Zh., Weiyan K. et al.//Sol.St.Communs. 1987.V.64.P.885.
11. Boiko B.B., Akimov A.I., Gatalskaya V.I. et al.// Izv. AN BSSR.1988.N5.P.47.

SUPERCONDUCTIVITY IN THE Bi-Sr-Ca-Cu-O AND
Tl-Ba-Ca-Cu-O SYSTEMS

M.P.PETROV, A.T.BURKOV, T.B.ZHUKOVA, A.V.IVANOV,
N.F.KARTENKO, M.V.KRASINKOVA, A.A.NECHITAILOV,
I.V.PLESHAKOV, V.V.PROKOFIEV, S.V.RAZUMOV, A.I.
GRACHEV, V.V.POBORCHY, S.S.RUVIMOV, S.I.SHAGIN,
V.V.POTAPOV, A.V.GOLUBKOV, A.V.PROCHOROV

A.F.Ioffe Physical Technical Institute of the
Academy of Sciences, Leningrad, 194021, USSR

ABSTRACT

Conditions for preparing single-phase ceramics with molar ratio 4:3:3:4 of the Bi-Sr-Ca-Cu-O sytem with $T_c(R=0)$ near 85 K have been found. The electrical, magnetic, and optical studies of ceramics and single crystals have been performed. Superconductivity with $T_c(R=0) = 100$ K has been observed in the Tl-Ba-Ca-Cu-O system. The superconducting phase has been identified. Specific features of synthesis of superconducting compounds of these systems are compared.

INTRODUCTION

The discovery of superconductivity at 30 K and then at 90 K in cuprates containing rare earth elements /1,2/ gave impetus to a wide search for new compounds with higher temperatures of transition into the superconducting state. These compounds proved to be also cuprates but free of rare earth elements and containing bismuth or thallium. As preliminary data indicate, in addition to higher T_c these compounds have the advantage of being less susceptible to environmental conditions (humidity, thermocycling) /3/. They are also more easily fabricated /4/ since there is no need for an additional annealing in oxygen at a temperature below that of synthesis. However, difficulties are encountered in preparation of single-phase samples of these compounds.

This paper discusses specific features of synthesis of Bi- and Tl-containing compounds and shows, in particular, how conditions of annealing in oxygen affect superconducting properties.

SYNTHESIS OF COMPOUNDS, THEIR STRUCTURES AND COMPOSITION

Compounds of the Bi-Sr-Ca-Cu-O system were prepared by two techniques, i.e., solid-state sinter reaction and spontaneous crystallization from the melt. The starting materials were Bi_2O_3, CuO, $SrCO_3$, and $CaCO_3$, each 99.99% pure. They were carefully mixed and grounded in a jasper montar with ethanol, then dried, placed into an alumina or Pt crucible and sintered at the temperature close to the melting point in case of solid-state reaction (as in /5/) and at T = 1000-1200°C in case of crystallization from the melt (as in growing single crystals /6/). The time needed for synthesis varied from several hours to dozens of hours. In order to accelerate the solid-state synthesis of the multicomponent system and to obtain a homogeneous material, the mixture was intermediately regrounded. In spontaneous crystallization, the melt was periodically mixed. The solid-state synthesis was carried out in air or oxygen flow, crystallization was performed only in air. In order to prepare samples in the form of pellets 1 mm thick and 10 mm in diameter, we used cold pressing under a pressure of $\simeq$ 2 tons/cm^2. The annealing cycle was repeated after pressing. In crystallization from the melt, samples were splitted or cut off by a diamond disk as parallelepipeds 2 x 3 x 10 mm^3 in size. When the melt was crushed, separate crystallites were found. They were both in the form of thin flakes with an area of $\simeq$ 1 mm^2 and $\simeq$ 10 μm thick and in the form of faceted bulk crystallites with one or several mirror faces, about 10 mm^3 in size.

The technique of synthesis of Tl-Ba-Ca-Cu-O compounds will be discussed below.

The phase composition of prepared pressed pellets and powders was determined by an X-ray diffractometer DRON-3 using Co- and Cu-K$_\alpha$ radiation. The X-ray powder diffraction patterns were recorded over the 2θ range from 5° to 90°.

The chemical composition of the synthesized compounds was determined by a chemical analysis. The content of bismuth, thallium and copper in the Bi-Sr-Ca-Cu-O and Tl-Ba-Ca-Cu-O systems was found by chelatometry titration /7/ at pH = = 1.5-2.0 for Bi and pH = 5.5-6.0 for Tl and Cu. The content of alkaline earth elements Ba, Sr, Ca was defined in ammoniac solution by titration of an EDTA excess using $MgCl_2$ after Bi, Cu and Tl were separated as diethildithiocarbominates. All the above analyses were performed using 0.1-0.2 grams of the substance studied.

The content of nonstoichiometric oxygen was found by the volumetrical technique /8/ using a 10-cm^3 volumeter to measure a small volume of gas. The sample studied was dissolved in the diluted chloric or nitric acid, with an accompanying isolation of the nonstoichiometric oxygen. For 0.4-2.0 weight % of nonstoichiometric oxygen, the variation coefficient was not higher than 5%.

THE Bi-Sr-Ca-Cu-O SYSTEM

The starting composition with a molar ratio of Bi:Sr:Ca:Cu = 4:3:3:4 allowed ceramics close to the single-phase one to be prepared. As X-ray powder diffraction revealed, this material has a layered perovskite structure with lattice constants a = 5.410(5) Å, b = 5.430(5) Å, and c = 30.81(4) Å. A small amount of phase with c = 24.5 Å and traces of CuO were also observed (Fig.1a). The samples with a less perfectly formed phase had a = 5.40 Å, b = 5.38 Å, and c = 30.72 Å.

The results of the chemical analysis are summarized in Table I.

It is seen that the resulting composition remains close to the starting one during synthesis and the content of nonstoichiometric oxygen is x = 0.54. This means that while the oxidation state of bismuth is 3, the formal oxidation state

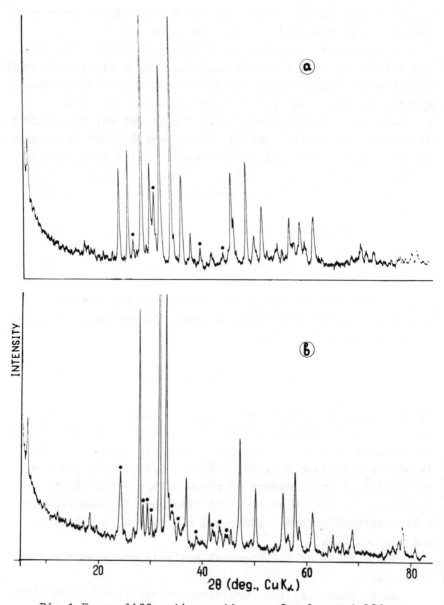

Fig.1 X-ray diffraction patterns of polycrystalline samples:
a - nominal composition $Bi_4Sr_3Ca_3Cu_4O_{16+x}$
b - nominal composition $Tl_2Ba_2Ca_2Cu_3O_{10+x}$
Points show reflections from nonsuperconducting phases.

of copper is 2.27[*], i.e., as in $Y_1Ba_2Cu_3O_{7-x}$. Since ceramics with the above mentioned component ratio was close to the single-phase one, it was used for studies of the effect of annealing conditions on superconducting properties. The best results were obtained for annealing at temperature of 860°C close to the melting one (as in /5/) and for rapid cooling after annealing.

TABLE I

Starting composition	Element	Content in the sample weight %	Standard deviation	Resulting composition
$Bi_4Sr_3Ca_3Cu_4O_{16+x}$	Bi	48.7	0.160	$Bi_{4.0}Sr_{2.66}-$
	Cu	14.8	0.0897	$Ca_{3.21}Cu_{4.00}-$
	Ca	7.5	0.143	$O_{15.87+0.54}$
	Sr	13.6	0.326	
	O_x	0.50	0.0212	

The temperature dependence of resistivity for these conditions is shown in Fig.2 (curve a). Resistivity was measured by the dc standard four-probe technique using indium contacts. Resistivity of the sample which was $\rho = 3$ mohm·cm at room temperature decreased linearly with decreasing temperature to ~ 115 K where a drop in resistivity was observed. As seen from Fig.2, the major change in resistivity occurs near $T_{c_{onset}}$ = 90 K and zero resistance is achieved at $T_{c_{R=0}}$ = 83 K.

In order to detect the diamagnetic signal and estimate the amount of the superconducting phase in the sample, we measured the high-frequency susceptibility. A coil with an internal volume approximately equal to the volume of the sample studied and an inductance equal to 500 μH was used. $\Delta L(T)$

---

[*]This is an approximate estimate since it does not take into account the content of the phase with c = 25.4, where copper is in the oxidation state 2 and possible presence of bismuth in the oxidation state 5.

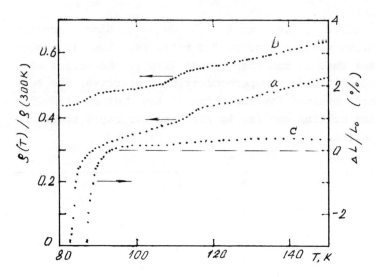

Fig. 2 Resistivity as a function of temperature for samples with nominal composition $Bi_4Sr_3Ca_3Cu_4O_{16+x}$
  a - after annealing in O_2 at a temperature close to the melting point followed by rapid cooling
  b - after long annealing (72 hours) in O_2 at 500-300 K.

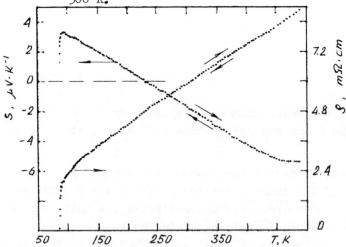

Fig. 3 Temperature dependence of thermoelectric power and resitivity for a sample with composition $Bi_4Sr_3Ca_3Cu_4O_{16+x}$.

which is proportional to the real part of magnetic susceptibility $\chi'(\omega)$ was measured at $\omega = 150$ kHz. The field inside the coil was less than 1 Oe. Estimates of the skin-effect contribution into magnetic susceptibility for a given conductivity of samples showed that it can be ignored. The error in measurements of $\Delta L/L_0$ is estimated to be $\sim 0.1\%^*$, with due account for temperature variations of inductance of the empty coil.

Fig.2 shows also the temperature dependence of $\Delta L/L_0$ for the same sample. It is seen that the resistivity drop at $T \simeq 115$ K is accompanied by a well-pronounced change of magnetic susceptibility (a low diamagnetic signal arises), which confirms the presence of superconductivity in a fraction of the sample volume with $T_c \simeq 115$ K. The major fraction of the sample volume passes into the superconducting state below 90 K. Fig.2c shows only the onset of the transition.

Measurements of the temperature dependence of thermoelectric power of one of the $Bi_4Sr_3Ca_3Cu_4O_{16+x}$ samples, with simultaneous measurements of resistivity between the same point contacts, are presented in Fig.3. It shows the cooling-heating-cooling cycle in high-purity helium. The thermoelectric power is linearly related to temperature. It has a low magnitude, positive sign near transition to the superconducting state and changes its sign with increasing temperature at $T \simeq 230$ K. At $T \simeq 110$ K, the $S(T)$ dependence deviates from linearity, this fact being associated, as noted above, with superconductivity in a small fraction of the sample. The phase responsible for superconductivity at this temperature has not yet been identified. This is likely to be due to its small amount in the samples studied.

The experiments involving different conditions of annealing in oxygen indicated that annealing for a long period of

---

*Variation of inductance ΔL is proportional not only to χ', but also to the coefficient of coil filling, whose exact value is difficult to measure. In this paper it was taken to be 50%.

time at temperatures somewhat lower than 860°C, as well as slow cooling after annealing in the O_2 atmosphere, make the superconducting properties poorer, i.e., a fraction of the superconducting phase in the sample decreases, the transition region becomes wider, and the temperature at which zero resistance is achived drops sharply. Fig.2 (curve b) shows for comparison the temperature dependence of resistivity for the sample which was annealed for a long time (72 hours) in O_2 at 500-300°C. It is seen that the superconducting state does not manifest itself at least to 77 K. From the X-ray analysis, no appreciable changes of the phase composition occur during annealing, the phase with c = 30.8 Å is preserved, though traces of CuO, $CaCuO_2$, and Ca_2CuO_3 impurities, which point to dissociation of the compound, are found. As the chemical analysis revealed, a long annealing in oxygen even at high temperatures does not affect the content of nonstoichiometric oxygen (at least within 0.02 weight %) if the phase with c = 30.8 Å is already formed. Thus, the behaviour of $Bi_4Sr_3Ca_3Cu_4O_{16+x}$ compound during annealing in O_2 differs from that of $YBa_2Cu_3O_{7-x}$, for which high T_c is achived only after long-term annealing at relatively low temperatures (450-550°C).

Summarizing the results, it should be noted that a single-phase compound of $Bi_4Sr_3Ca_3Cu_4O_{16+x}$ with c = 30.81 Å passes into the superconducting state with $T_{c,onset}$ = 90 K only after annealing in O_2 at temperatures close to the melting one, with a rapid cooling to follow. Annealing does not affect the content of nonstoichiometric oxygen in this case. The obtained data can be explained by a fine balance between the oxidation state of Cu which is in the intermediate valence state Cu^{3+} and Cu^{2+} and oxidation state of bismuth Bi^{3+} and Bi^{5+}. Indeed, Bi^{5+} can be obtained in this system only at high temperatures in the presence of oxygen and can be preserved by rapid cooling. It is possible that the effect of Bi in this system is similar to that in $BaPb_{1-x}Bi_xO_3$, where simultaneous presence of Bi^{3+} and Bi^{5+} stimulates pairing of carriers.

In order to identify phases in the Bi-Sr-Ca-Cu-O system,

crystallites obtained by spontaneous crystallization of melts of different compositions were studied.

The X-ray analysis of the crystallites and single crystals obtained reveals the presence of at least two phases of a layered perovskite structure with parameters $a = 5.410(5)$, $b = 5a = 27.00(5)$, and $c = 30.90(6)$ Å, and also with $c = 24.52(6)$ Å. The probe microanalysis identified the composition of these phases as $Bi_2Sr_{2.2}Ca_{0.8}Cu_2O_x$ and $Bi_2Sr_{1.25}Ca_{0.25}CuO_x$, which is consistent with the results reported in /9,10/.

Complementary data on the structure of the material were obtained in the transmission electron microscopy study of thin crystal plates $0.3 \times 0.3 \times 0.01$ mm^3 in size which undergone auxiliary ion etching. The plates were fixed on a mesh using indium or gallium and then thinned from both sides by an Ar^+ ion beam (5 kV, 30 μA) for several hours. During etching, samples gave a rose-blue luminescence. Thin transparent rose regions were studied in a transmission electron microscope JEM-7A with an accelerating voltage of 100 kV.

Fig.4 shows a typical electron diffraction pattern for the [001] orientation of the sample. It is seen that the crystal is orthorhombic, with $a = 5.4$ Å and $b = 5a = 27$ Å. A more accurate value of b/a is about 4.8, which is likely to be indicative of the presence of an incommensurate phase with the perovskite-related subcell in this crystal. Satellites on the pattern in the [010] direction which correspond to superstructure may be attributed to both ordering of solid solutions of impurities and vacancies, alternation of Sr and Ca ions, and formation of an incommensurate phase due to displacement of Bi ions from the lattice sites.

On the whole, the samples studied were sufficiently homogeneous and contained a small amount of other phases, lamellae and regions with other orientations of crystallographic axes.

Measurements of temperature dependences of resistivity and magnetic susceptibility of polycrystalline agglomerates indicated that samples having a dominating phase with $c =$

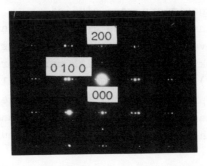

Fig. 4 Electron diffraction pattern of a superconducting crystal of the Bi-Sr-Ca-Cu-O system corresponding to the c zone axis. Indexes of diffraction spots correspond to an ideal superlattice with b = 5a, though the real superlattice is incommensurable.

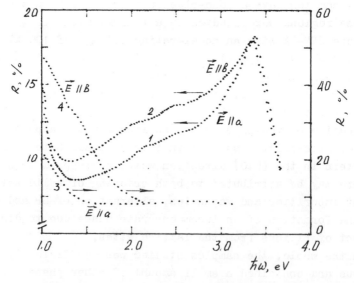

Fig. 5 Reflectivity spectra of superconducting crystals of the Bi-Sr-Ca-Cu-O system (1,2) and $YBa_2Cu_3O_{7-x}$ (3,4) in the a-b plane for two orthogonal polarizations of a light wave (E is the electric field strength vector of the light wave).

30.9 Å are superconducting, with the onset of transition into the superconducting phase at $T \simeq 90$ K and zero resistance at 83 K, as in ceramic samples described above.

Single-crystal plates (or perfect faces of crystallites) were optically studied (similar to /11/) in a polarizing microscope. Their optical anisotropy in the a-b plane is well seen in the reflected light. In contrast to /11/, twins with two types of boundaries, i.e., along (100) and (110), were observed. The single-domain regions were typically 10-100 μm in size.

Fig.5 shows the reflectivity spectra for the a-b plane of a single-domain region in an energy range $\hbar\omega = (1-3.6)$ eV. Specific features of the spectra are the peak at $\hbar\omega = 3.35$ eV and reflectivity rise at $\hbar\omega \simeq 1.3$ eV. According to /12/, the latter is attributable to the plasma edge. Anisotropy of optical properties is also well seen in the spectra.

The optical reflectivity anisotropy in the a-b plane was also observed in $YBa_2Cu_3O_{7-x}$ crystals /11/ (curves 3 and 4 in Fig.5). It was interpreted as an anisotropy of the high-frequency (optical) conductivity which is due to the presence of Cu-O chains along the b axis. However, while the reflectivity anisotropy in $YBa_2Cu_3O_{7-x}$ manifests itself in the shift of the plasma edge for one of the orthogonal polarizations relative to the second one, for crystals of the Bi-Sr-Ca-Cu-O system it expresses itself only in quantitatively different reflectivities. In our opinion, this can indicate that the Bi_2O_2 layers which are responsible for the superstructure /13/ and, hence, the nonequivalence of the a and b axes does not play a significant role in charge transport, at least in the region of optical frequencies. This conclusion is consistent with a vast amount of data on the dominating role of CuO_2 layers in the mechanism of conductivity for all presently known high T_c superconductors.

THE Tl-Ba-Ca-Cu-O SYSTEM

The technique for preparation of the sample with nominal composition $Tl_2Ba_2Ca_2Cu_3O_{10+x}$ is similar to that reported in

/14,15/. The starting components were Tl_2O_3, $BaCuO_2$, and Ca_2CuO_3 with the molar ratio $Tl:Ba:Ca:Cu = 2:2:2:3$. The material was synthesized in air at 700°C, then pellets 10 mm in diameter and 1 mm thick were pressed and annealed in oxygen flow under different conditions. The best results were obtained, like in /15/, for a short annealing at $T > 800°C$ (when Tl^* begins to evaporate strongly) followed by rapid cooling.

The powder X-ray diffraction pattern of the sample with nominal composition 2:2:2:3 (Fig.1b) shows the main peaks corresponding to the 2:2:1:2 phase. This is a layered perovskite structure with lattice constants $a = b = 3.85 \pm 0.05$ Å and $c = 29.30 \pm 0.05$ Å.

The pressed pellets give also reflection from a nonidentified phase with parameter $c = 11.55$ Å (or 23.10 Å) (reflection 001). This phase is likely to be formed on the surface, it is strongly textured and this is the evidence of its layered structure. Only traces of this phase were observed in the powder obtained by grounding samples. Judging by the observed 001 reflections, this phase can be isostructural with $YBa_2Cu_3O_{7-x}$. Several extra peaks in the diffraction pattern (Fig.1b) can be attributed to some impurity phases, such as $Ba_{1-x}Ca_xCO_3$, $BaCaO_2$ and CuO.

The chemical analysis of the compound is summarized in Table II.

TABLE II

Starting composition	Element	Content in the sample, weight %	Standard deviation n = 5	Resulting composition
$Tl_2Ba_2Ca_2Cu_3O_{10+x}$	Cu	17.4	0.0552	$Tl_{1.91 \pm 0.01}$
	Tl	35.7	0.238	$Ba_{1.96 \pm 0.02}$
	Ba	24.2	0.309	$Ca_{1.97 \pm 0.03}$
	Ca	7.2	0.0951	$Cu_{3.00 \pm 0.009}$
	O_x	1.2	–	$O_{9.80 + 0.82}$

*Tl evaporates as Tl_2O

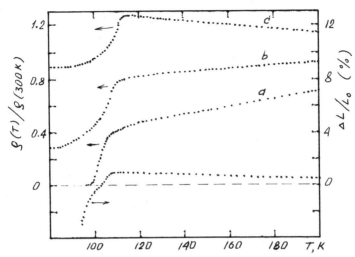

Fig.6 Resistivity as a function of temperature for a sample with nominal composition $Tl_2Ba_2Ca_2Cu_3O_{10+x}$
  a - after annealing in O_2 at the temperature at which evaporation begins (Tl)
  b - after annealing in O_2 at 400°C for 20 hours
  c - without annealing in O_2
Temperature dependence of $\Delta L/L_o$ for sample "a".

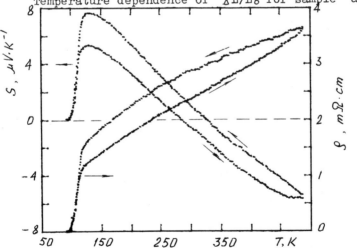

Fig.7 Temperature dependence of thermoelectric power and resistivity for the same sample as in Fig.6 (a heating-cooling cycle in He).

The analysis revealed that a small portion of Tl is lost during synthesis. The content of nonstoichiometric oxygen is 0.82 oxygen atom/formula unit. If we suppose that the oxidation state of Tl in this compound is 3 and neglect other Cu-containing phases, the formal oxidation state of copper is 2.55, which is well above that in $YBa_2Cu_3O_{7-x}$.

Fig.6a,b,c presents resistivity vs. temperature for different conditions of annealing in O_2. The curve for the non-superconducting state changes its behaviour from a semiconducting to metallic one with increasing annealing temperature. As evidenced by the X-ray analysis, no detectable changes in phase composition and structure are observed. Deviation from linearity of R(T) for sample "a" begins at T=120 K and may be regarded as the onset of superconducting transition. A sharp transition is at T = 100 K. Fig.6 shows also the temperature dependence of $\Delta L/L_0$ (which is proportional to χ'). The diamagnetic signal appears at $T \simeq 100$ K. The amount of the superconducting phase with $T_c \simeq 100$ K is about 3%.

Fig.7 presents the temperature dependence of thermoelectric power and resistivity measured simultaneously between the same points of the sample from 77 to 500 K in pure He. A heating cycle with subsequent cooling is shown. The transition region is fairly wide, with an onset at 120 K and zero resistance at 93 K. However, after heating in He $T_{c_{R=0}}$ decreases to 89 K and resistivity in the normal state increases. The thermoelectric power is positive near T_c and changes its sign at $T \simeq 280$ K. The thermopower of some samples was observed to decrease at higher temperatures (160 K), which was likely to be due to a presence of a small amount of phase with higher T_c. The amount of this phase was so small that no superconducting current passed through the sample.

Low values of thermopower and also sign inversion are typical of materials where concentrations of electrons and holes are nearly equal. Since the electron gas in CuO_2 layers is characterized by strong correlation, this must correspond to the intermediate valence state of copper Cu^{z+} with $z \simeq 2.5$, which is consistent with our experimental data on the content of nonstoichiometric oxygen.

Thus, judging by the thermopower, the band-filling pattern in Tl- and Bi-containing ceramics is similar to that of $YBa_2Cu_3O_{7-x}$ rather than of $(La_{1-x}Sr_x)_2CuO_4$ which has a hole conductivity with a fairly high thermopower ($\simeq 60$ μV degr^{-1}) /16/.

It is still rather difficult to explain why annealing at temperatures at which thallium oxide begins to decompose is necessary. It can only be supposed that the presence of Tl in two different oxidation states (3 and 1) is a factor of great importance for superconductivity in this system. The oxidation state 1+ can be obtained at $T > 800°C$. It is known /17/ that thallium oxide (where thallium oxidation state is 3+) begins to dissociate to monoxide where the oxidation state of Tl is 1+. The monoxide begins to evaporate intensively above 500°C. Rapid cooling allows this oxidation state to be preserved to room temperature. A slow cooling will lead to an inverse transition $Tl^{1+} \rightarrow Tl^{3+}$.

Thus the Bi-Sr-Ca-Cu-O and Tl-Ba-Ca-Cu-O systems are similar to some extent to each other as far as the presence of the element which changes its oxidation state by two units (Bi^{3+} and Bi^{5+}, Tl^{3+} and Tl^{1+}) and thereby creates pairs of carriers is concerned. It can be supposed that this stimulates pairing of carriers in CuO_2 layers responsible for superconductivity and can lead to an increase of T_c, similar to both $Pb_{1-x}Tl_xTe$ /18/ and $Ba_{1-x}K_xBiO_2$ /19/.

ACKNOWLEDGEMENTS

The authors would like to express their thanks to Dr. B.Ya.Moizhes for helpful discussions and to Drs. S.G.Konnikov and V.V.Tretyakov for performing the probe analysis of crystallites.

REFERENCES

1. Bednorz J.G. and Müller K.A. Z.Phys., 1986, B 64, S.189.

2. Wu M.K., Ashburn J.R., Torng C.J., Hor P.H., Meng R.L., Gao L., Huang Z.J., Wang Y.Q., and Chu C.W., Phys.Rev.Lett. 1987, v.58, p.908.
3. Nasu H., Makida S., Ibara Y., Kato T., Imura T., Osaka Y., Jpn.J.Appl.Phys., 1988, v.27, p.L536.
4. Maeda H., Tanaka Y., Fukutomi M., Asano T., Jpn.J.Appl. Phys., 1988, v.27, p.L209; Komatsu T., Imai K., Sato R., Matusita K., Yamashita T., Jpn.J.Appl.Phys., 1988, v.27, p.L533.
5. Tallon J.L., Buckley R.G., Gilberd P.W., Presland M.R., Brown I.W.M., Bowden M.E., Christian L.A., Goguel R., Nature, 1988, v.333, p.153.
6. Liu J.Z., Grabtree G.W., Rehn L.E., Geiser U., Young D.A., Kwok W.K., Baldo P.M., Williams J.M., Lam D.J., Phys.Lett. A, 1988, v.127, p.444.
7. Schwartzenbach G., Flashka G., Chelatometry Titration, Moscow, "Chimia", 1970, 359 p.
8. Samsonov G.V., Pilipenko A.T., Hazarchuk T.N., Analysis of Refractory Compounds, Moscow, "Mettallurgizdat", 1962, 103 p.
9. Chippendale A.M., Hibble S.J., Hriljac J.A., Cowey L., Bagguley D.M.S., Day P., Cheetham A.K., Physica C, 1988, v.152, p.154.
10. Takayama-Muromachi E., Uchida Y., Oho A., Izumi T., Onoda M., Matsui Y., Kosuda K., Takekawa S., Kato K., Jpn.J. Appl.Phys., 1988, v.27, p.L365.
11. Petrov M.P., Grachev A.I., Krasinkova M.V., Nechitailov A.A., Prokofiev V.V., Poborchy V.V., Shagin S.I., Kartenko N.F., Pis'ma v Zh.Tekh.Fiz., 1988, v.14, p.748.
12. Takagi H., Eisaki H., Uchida S., Maeda A., Tajima S., Uchinokura K., and Tanaka S., Nature, 1988, v.332, p.236.
13. Matsui Y., Maeda H., Tanaka S., Horiuchi S., Jpn.J.Appl. Phys., 1988, v.27, p.L372.
14. Kikuchi M., Kobayashi N., Iwasaki H., Shindo D., Oku T., Fokiwa A., Kajitani T., Hiraga K., Syono Y., Muto Y., Jpn.J.Appl.Phys., 1988, v.27, p.L1050.
15. Sheng Z.Z., Hermann A.M., Nature, 1988, v.332, p.138.

16. Yu R.C., Naughton M.J., Yan X., and Chaikin P.M., Holtzberg F., Greene R.L., Stuart J., Davies P., Phys.Rev. B, 1988, v.37, p.7963.
17. Chemistry and Technology of Rare-Earth Elements, Ed. by Bolshakov K.A., Moscow, "Vysshaya Shkola", 1978, 367 p.
18. Chernik I.A., Lykov S.N., Fiz.Tverd.Tela, 1981, v.23, p.1400.
19. Cava R.J., Batlogg B., Krajewski J.J., Farrow R., Rupp Jr. L.W., White A.E., Short K., Peck W.F., and Kometani T., Nature, 1988, v.332, p.814.

ON POSSIBLE METHODS OF SYNTHESIS AND PROPERTIES OF SUPERCONDUCTIVE THALLIUM CUPRATES

Ozhogin V.I., Shustov L.D., Myasoedov A.B., Tolmacheva N.S.,
Inyushkin A.V., Taldenkov A.N., Babushkina N.A.,
Krasnoperov E.P. Teterin Yu.A.

Kurchatov Institute of Atomic Energy, 123182, Moscow, USSR

Abstract

Some methods of synthesis of metal-oxide HTSC ceramics based on the Tl-Ca-Ba-Cu-O composition and methods of improvement of their quality have been investigated. The dependence of characteristics of HTSC samples on temperature and heat treatment duration has been studied for a mixture of $Ba_2Ca_3O_5$ (or $Ca_2Ba_2Cu_3O_7$) and Tl_2O_3 and optimum conditions for preparation of the samples have been found. Samples with T_c = 103 and 117 K were prepared by three-stage heat treatment of a mixture of Tl_2O_3, CaO, $Ba(NO_3)_2$ and CuO. The temperature dependences of their resistance $R(T)$ and magnetic susceptibility $\chi(T)$ have been determined. X-ray phase analysis of the samples has been made and the dependence of their resistance on magnetic field has been investigated.

The XPC data for the HTSC samples showed that the oxidation state of copper ions is close to Cu(1) and Cu(2), that of thallium ions to Tl(1); calcium and barium exist in two oxidation states, one of which is essentially lower than Ca(2) and Ba(2), respectively. The fine structure of XPS spectra in the energy range of 10-40 eV is assumed to result from the interaction of Tl5d-, Ca3p-, Ba5p-, Ba5s- and O2s- shells of the neighbouring atoms. This structure is an evidence of affecting the inner valency molecular orbitals to the state of the outer-shell electrons.

A scheme of the structure of the thallium-based metal-oxide HTSC ceramics is proposed with the XPA and XPS data taken into consideration.

The attempts of researchers working on synthesis of high-temperature superconductors (HTSC) to prepare ceramics with T_c above 100 K have led to the discovery of a new class of HTSC on the basis of the thallium-calcium-barium-copper-oxygen composition [1-3]. Despite the absence of acknowledged theoretical grounds in synthesis of new superconductive structures, which was performed to date empirically, some regularities concerning the ionic radius and the valencies of elements were revealed in the course of substituting successively ittrium in the $YBa_2Cu_3O_{7-x}$ system by other rare earths. The substitution of thallium for ittrium led to synthesis of a superconductive Tl-Ba-Cu-O system with T_c = 80 K; the introduction of calcium into this system made it possible to increase T_c to 105 K and 125 K.

Some possible methods of synthesizing metal-oxide HTSC ceramics based on the Tl-Ca-Ba-Cu-O system and improving their quality have been investigated in this work; their structure and properties have been also examined. The temperature dependences of resistance R(T) and magnetic susceptibility $\chi(T)$, and the dependence of resistance on magnetic field have been measured in the samples obtained; X-ray phase analysis (XPA) has been made; XPS spectra of the samples have been also taken. A scheme of the elementary cell of the metal-oxide HTSC ceramics based on the Tl-Ca-Ba-Cu-O composition was proposed from the consideration of the results obtained.

Analytically-pure thallium, calcium and copper oxides and barium carbonate and spectrally-pure barium nitrate were used as source reagents to synthesize the superconductive samples. All stages of synthesis were made in platinum crucibles placed into a quartz tube heated by an electric furnace in flowing dried oxygen. The preparation of the source and intermediate mixtures at each stage of synthesis was accompanied by their homogenization. The samples obtained represented black-colored pellets 12 mm in diameter and 0.8-1.5 mm thick.

The values of T_c for the samples were determined both resistively using the installation [4] and by control of appearance of a diamagnetic signal in a magnetometer. The ac susceptibility was measured in a magnetic field of 1 oersted at a frequency of 667 Hz. The onset temperature was determined by the intersection of the extrapolated temperature dependences of normal resistance and superconducting transition itself.

The specificity of the thallium-based HTSC ceramics results not so much from toxicity of thallium oxides (Tl_2O, Tl_2O_3, Tl_4O_3) as from their relatively high volatility accompanied by oxidation and dissociation of products (the melting point of Tl_2O_3 is 717°C). Therefore, to minimize decomposition and volatilization of thallium oxides the procedure of synthesis of superconductive Tl-based materials must ensure high temperatures to initiate a superconductive structure for short times of heat treatment. Naturally, technologists are striving, using the conventional methods of solid-phase synthesis, primarily to obtain cuprates of elements constituting the superconductive composition and only at a final stage to introduce thallium oxide by heat treatment of short duration.

In reproducing one of the methods of synthesis of Tl-based HTSC ceramics described in [1,2] we added appropriate amounts of CaO and Tl_2O_3 to $Ba_2Cu_3O_5$ prepared by heat treatment of the $BaCO_3$-CuO mixture at 925°C for 28-39 hours. The mixture was pressed into pellets and treated for 2-6 min at 880-910°C in flowing oxygen. The samples were cooled either in air or in oxygen. In both cases the superconductive properties were similar. The variation of the initial Tl:Ca:Ba:Cu ratio almost always resulted to formation of superconductive materials with different T_c.

We have investigated the effects of temperature and time of final heat treatment within the above limits (880-910°C, 2-6 min) on the characteristics of the samples obtained at the same initial ratio Tl:Ca:Ba:Cu = 2:2:2:3. It was experimentally shown that practically there is no systematic dependence of T_c and ΔT_c on heat treatment conditions. These parameters varied from 80 to 115 K with a superconducting transition wider than 10 K. Figure 1 shows the temperature dependence of susceptibility for a typical sample prepared by the procedure described (curve 1).

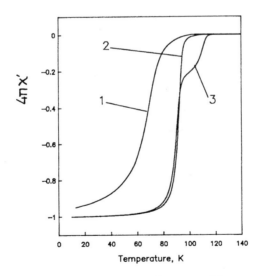

Fig. 1. Temperature dependence of magnetic susceptibility for $Tl_2Ca_2Ba_2Cu_3O_{10+x}$ pellets. The heating temperature and time: 1 - 875°C, 5 min; 2 - 895°C, 5 min; 3 - 910°C, 4 min.

A high reproducibility of synthesis of samples with T_c=105 K and ΔT_c = 5 K (curve 2, fig. 1) was reached in using cuprate $Ca_2Ba_2Cu_3O_7$ obtained by heat treatment of the $CaO-BaCO_3-CuO$ mixture in air for 30 h at 925°C followed by addition of Tl_2O_3 and heat treatment of the pellets for 5 min at 895°C. Increasing temperature of final heat treatment to 910°C promoted the enrichment of the samples with a superconductive phase with T_c = 117 K (curve 3, fig. 1). A further oxydizing heat treatment at 680-800°C for several hours improved essentially the quality of such samples with a maximum weight loss of less than 2 wt% (fig. 2).

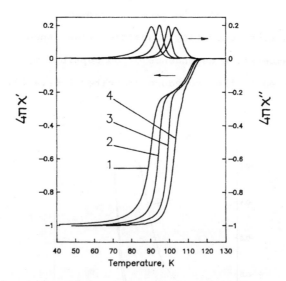

Fig. 2. Temperature dependence of real (χ') and imaginary (χ'') parts of magnetic susceptibility for the $Tl_2Ca_2Ba_2Cu_3O_{10+x}$ sample (910°C, 4 min) after its additional isothermal heat treatment in flowing oxygen at various treatment temperatures and times: 1 - original sample; 2 - 680°C, 2 hours; 3 - 710°C, 4 hours; 4 - 800°, 3 hours.

The quality of the samples synthesized depends not only on the presence of some superconductive phase, but also on the presence of $BaCO_3$ due to its incomplete fixation at the stages of synthesizing source cuprates $Ba_2Cu_3O_5$ or $Ca_2Ba_2Cu_3O_7$. $BaCO_3$ has a high point of thermal decomposition (1450 C in air) which naturally decreases in the reaction mixture with CuO and CaO; nevertheless its complete fixation is not, as a rule, reached.

In order to overcome this difficulty we proposed a method of synthesis of Tl-based HTSC ceramics from the mixture of Tl_2O_3, CaO, $Ba(NO_3)_2$ and CuO. The method represents a three-stage heat treatment of the mixture in flowing oxygen. At the first stage the mixture was heated from $20°$ to $720°C$ with a rate of 360 C/h, whereby $Ba(NO_3)_2$ was decomposed to BaO. The second high-temperature stage required to form a superconductive structure consisted in an isothermic treatment at $900°C$ for 30 min. The losses of Tl_2O_3 (to 40-45 %) at this stage were compensated by an excess initial amount of this oxide over stoichiometry. The final stage represented a short-time heat treatment of the pressed pellets at $890°C$ for 5 min followed by cooling the sample in flowing oxygen. Superconductive samples with T_c = 103 K and ΔT_c = 6 K (103 K sample), and T_c = 117 K and ΔT_c = 4 K (117 K sample) were synthesized by this method. Figures 3 and 4 present the dependences R(T) and χ(T) for these samples. The Tl:Ca:Ba:Cu ratio was 4:1:1.5:1.4 in the source mixture and 2:1:1.5:1.4 in the final HTSC sample. As follows from the results presented, materials containing different fractions of superconducting phases were obtained from the source mixtures of the same composition under identical conditions. Such a phenomenon seems to result from a proximity of the structural and kinetic parameters of superconducting phase formation, which requires a further systematic study.

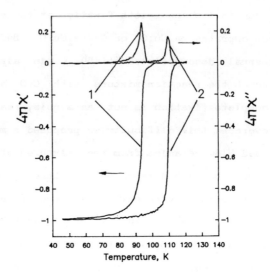

Fig. 3. Temperature dependence of real (χ') and imaginary (χ'') parts of magnetic susceptibility for the samples with T_c = 103 K (1) and 117 K (2).

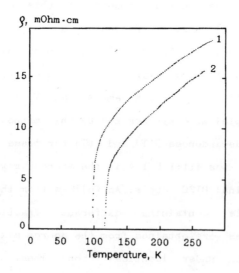

Fig. 4. Temperature dependence of resistance for the samples with T_c = 103 K (1) and 117 K (2).

In synthesizing the Tl-based HTSC ceramics we used, as source reagents, thallium, calcium, barium and copper perxenates: $Tl_4(XeO_6)_3$, Ca_2XeO_6, Ba_2XeO_6 and Cu_2XeO_6 in appropriate proportion. These perxenates are easily decomposed ($t < 400°C$) to corresponding oxides to liberate elemental xenon and oxygen by the reactions:

$$Tl_4 (XeO_6)_3 \longrightarrow 2 Tl_2O_3 + 3Xe + 6O_2,$$
$$M_n XeO_{4+n} \longrightarrow nMO + Xe + 2O_2,$$

where M is the two-valent cation. The thermal decomposition of mixed perxenates was made by slow heating from $20°$ to $400°C$ with a rate of $100°C/hr$ in flowing oxygen. Superconductive samples were obtained after homogenization of the mixture, pressing and heat treatment of the pellets at $895°C$ for 5 min.

The measurement of the resistance versus magnetic field, XPA and XPS were carried out on the 103 K and 117 K samples; the characteristics of these samples are shown in figs. 3 and 4.

Figure 5 gives the dependence of resistance on magnetic field for the 117 K sample; the values of resistance are normalized to the value of normal resistance R_n (at $T = 118$ K). It is seen that the resistance grows as the magnetic field increases, this dependence being linear at low temperatures and in strong fields. Such a behaviour of $R(H,T)$ gives good reasons to believe that due to very weak pinning there exists a dynamic resistive state caused by motion of the magnetic vortex structure. This circumstance makes it difficult to choose the criterion for determining the boundary of the phase transition (H_{c_2}) from the superconducting state to the normal one near the transition end. The value of dH_{c_2}/dT which turned out to be 2.2

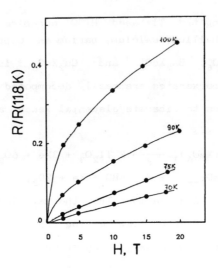

Fig. 5. Dependence of resistance on magnetic field for the sample with $T_c = 117°K$.

T/K was determined by the displacement of the transition onset in magnetic field. This value is typical of HTSC series. In the vicinity of the transition end point (R=0) the determination of H_{c_2} is a matter of some difficulties.

The XPA of the samples revealed the presence of a tetragonal structure with lattice constants: $a = b = 5.453 ± 0.001 Å$, $c = 29.80 ± 0.02 Å$ for the 103 K sample and $a = b = 5.453 ± 0.001 Å$, $c = 35.60 ± 0.02 Å$ for the 117 K sample. The authors of [1-3] which obtained thallium-based HTSC ceramics reported about the existance of three superconducting phases with T_c = 80 K, 105 K and 125 K and the Tl:Ca:Ba:Cu ratio equal to 2021, 2122 and 2223 respectively. Our samples with T_c = 103 K contained about 80 % of phase 2122, the samples with T_c = 117 K - about 80 % of phase 2223 and about 15 % of phase 2122.

The XPS spectra of the samples were measured using ar

electrostatic spectrometer HP 5950 A and exciting monochromatized Al K$\alpha_{1,2}$ (1486.6 eV) X-rays in vacuum at 1.3×10^{-7} Pa. The surfaces of the bulk ($10 \times 6 \times 2$ mm^3) samples bonded with cyacrin on aluminium substrates were mechanically cleaned by a steel scraper in vacuum at 10^{-4} Pa. The electron binding energies are given in the work relative to E_B(C1s) = 285.0 eV, the C1s binding energy for hydrocarbon vapours condensed on the sample surface. The binding energy of carbon on a gold platee is E_B(C1s) = 284.7 eV at E_B(Au 4f$_{7/2}$) = 83.8 eV. The errors in determining the electron binding energies and the relative intensities of lines were ± 0.1 eV and ± 10 % respectively.

The mechanical cleaning of the surface of the Tl-based HTSC ceramic samples caused essential changes in the structure of the inner and outer electron spectra. This is mainly due to the differences in stoichiometry and degree of ion oxidation on the surface and in the bulk. It follows from comparison of the spectra that the degree of oxidation of barium ions on the uncleaned surface is essentially close to Ba(2) (fig. 6). On the cleaned surface there exist at least two types of barium ions differing in degree of oxidation. Thus, the difference between the Ba 4d$_{5/2}$ binding energies for these ions equals to ΔE_B(Ba 4d$_{5/2}$) = 1.2 eV. The Ba 4d$_{5/2}$ binding energy for the high-energy component is comparable to the value E_B(Ba 4d$_{5/2}$) = 89.1 eV for BaCO$_3$ and it can be assigned to ions Ba(2) (see the table).

Table. Binding energies E (eV) of electrons.

	Subst.	$Tl_2Ca_1Ba_2Cu_2O_{8+x}$	Tl_2O	$CaCO_3$	$CaZrO_3$	$BaCO_3$	$BaZrO_3$	Cu_2O	CuO
MO	Mnlj								
EMO		-2.1							
		3.5	2.9	5.1	5.6	2.9	4.0	2.9	4.6
				10.0		5.1	5.3	6.2	12.3
				11.7					
IVMC	$Ba5p_{3/2}$	13.0				14.1	13.9		
	$Tl5d_{5/2}$	13.2	12.7						
	O2s	21.2	22.2	20.5	20.6		20.4	21.7	22.0
		23.7					22.0		
	Ca3p	23.4		25.0	24.8				
	Ba5s	27.8				29.0			
IMO	$Tl4f_{7/2}$	117.8	117.6						
	$Tl4f_{5/2}$	122.3	122.1	43.8	43.6				
	$Tl4d_{5/2}$	385.2	385.0	Ca3s	Ca3s		181.5		1097.6
	$Ca2p_{3/2}$	345.4		347.0	346.8				
	$Ca2p_{1/2}$			350.7	350.3				
	$Ba4d_{5/2}$	87.9				89.1	89.0		
	$Ba4d_{3/2}$			439.1		96.6	91.6		
	$Ba3d_{5/2}$	779.5		Ca2s					
	$Cu3p_{3/2}$							75.2	76.2
	$Cu2p_{3/2}$	932.7						932.6	933.6
	Sat	941.4						943.8	942.9
	$Cu2p_{1/2}$	952.7						952.3	953.3
	Sat	961.8						963.4	962.8
	O1s	528.9	528.9	532.0	529.9		529.5	530.9	530.0
		531.1				531.1			

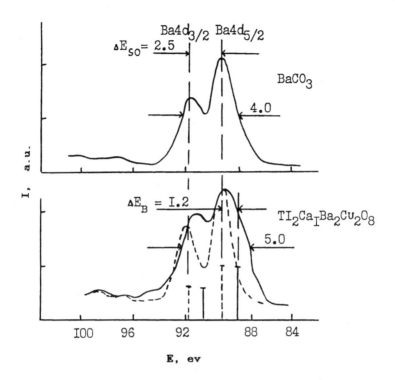

Fig. 6. Ba 4d XPS spectra for $BaCO_3$ and for sample with $T_c=103$ K: dotted line - uncleaned surface; solid line - surface scraped in vacuum. The linewidth is reduced to $\Gamma(C1s) = 1.3$ eV; no intensity calibration was made.

Decreasing the binding energy for barium ions of other type by ΔE_B ($Ba4d_{5/2}$) = 1.2 eV against the corresponding value for Ba(2) indicates that the degree of oxidation of these ions is less than two. Therefore, we designate formally these weakly-oxidized barium ions by Ba(0). An unambiguous connection between the fraction of Ba(0) in a sample and the behaviour of its superconductivity was earlier established on the basis of investigating a great number of the Y-Ba-Cu-O metal-oxide ceramic samples. Ba(O) free samples exhibit no HTSC behaviour.

Similar changes are observed in the Ca 2p spectra upon cleaning the sample surface (fig. 7). Thus, the difference between

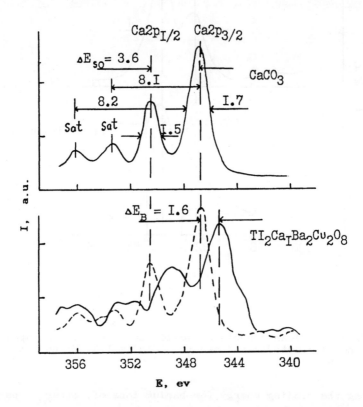

Fig. 7. Ca 2p XPS spectra for monocrystalline $CaCO_3$ and for the sample with T_c = 103 K: dotted line - uncleaned surface; solid line - surface scraped in vacuum. The linewidth is reduced to $\Gamma(C1s) = 1.3$ eV.

the Ca 2p binding energies for ions with different degrees of oxidation is $\Delta E_B(Ca\ 2p_{3/2}) = 1.6$ eV. A low-energy component at $E_B(Ca\ 2p_{3/2}) = 345.4$ eV was mainly observed after cleaning the surface. The Ca $2p_{3/2}$ binding energy on an uncleaned surface is close to its value $E_B(Ca\ 2p_{3/2}) = 347.0$ eV for $CaCO_3$ (see the table). By analogy with Ba(O) we designate formally the weakly-oxidized ions of calcium (E_B(Ca $2p_{3/2}$) = 345.4 eV) by Ca(O).

The degree of oxidation of thallium in the bulk of samples is less than that in Tl_2O_3. The binding energy of Tl $4f_{7/2}$ electrons is E_B(Tl $4f_{7/2}$) = 118.6 eV for Tl(3) in Tl_2O_3 and 117.6 eV for Tl(1) in Tl_2O (see the table). However, a "shoulder" observed on the high-energy side of the Tl 4f peaks characterizes the presence of Tl(3) ions in the sample.

The Cu 2p spectrum for a HTSC sample (fig. 8) differs essentially in its behaviour from that for CuO containing Cu(2)

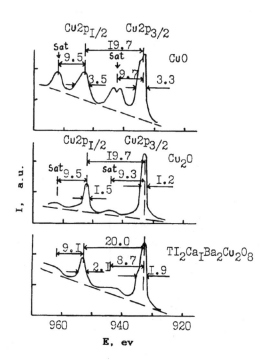

Fig. 8. Cu 2p XPS spectra for copper oxides CuO, Cu_2O and for the sample with T_c=103 K. The linewidth is reduced to Γ(C1s)=1.3 eV.

[5] and is closer to that for Cu_2O with $Cu(1)$. Indeed, the Cu 2p spectrum for the samples under study is representative of a mixture of copper ions of at least two degrees of oxidation: $Cu(1)$ and $Cu(2)$. It follows from an estimate based on the Cu $2p_{3/2}$ spectra that in the sample studied $Cu(1)$ and $Cu(2)$ account for 60 % and 40 % of copper ions, respectively. This estimate was made in the approximation that no "shake-up" satellites arise in the Cu 2p spectra for $Cu(1)$ with electron configuration $\{3\ d^{10}\}$ in contrast to the corresponding spectrum for $Cu(2)$.

The O 1s spectrum measured on an uncleaned surface has a complex shape due to surface oxidation of metals contained in a HTSC sample and adsorption of the oxygen-containing molecules (fig.9).

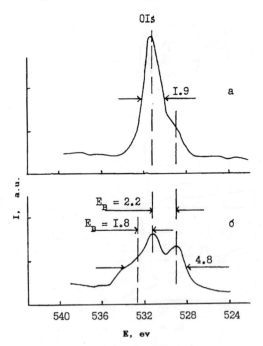

Fig. 9. O 1s XPS spectra for the sample with T_c = 103 K: a - uncleaned surface; b - surface scraped in vacuum. The linewidth is reduced to $\Gamma(C1s)$ = 1.3 eV.

Several peaks were observed in the O 1s spectrum after cleaning the sample surface. It follows from the results of extensive study of the O 1s spectra that the intensity of the low-energy component $E_B'(O\ 1s) = 528.9$ eV is more stable both in time and in passing from one sample to another in contrast to the corresponding values of the high-energy components $E_B''(O\ 1s) = 532.7$ eV. It was shown that the dwell of a sample in the chamber of the spectrometer for several hours leads only to a small increase in the intensity of the high-energy components in the O 1s spectrum.

To estimate the length of the metal-oxygen bond R_{M-O} from the O 1s spectra it is possible to use the dependence of the binding energy E_B (O 1s) on $1/R_{M-O}$ [6]. It is not difficult to find that $R_{M-O}' = 0.244$ nm corresponds to $E_B'(O\ 1s) = 528.9$ eV, $R_{M-O}'' = 0.196$ nm to $E_B''(O\ 1s) = 531.1$ eV and $R_{M-O}''' = 0.169$ nm to $E_B'''(O1s) = 532.7$ eV. It follows from these data that if two low-energy components of the O 1s spectrum can be assigned to oxygen atoms bound with ions of copper (E_B'', R_{M-O}''), thallium, barium and calcium (E_B', R_{M-O}'), then the high-energy component (E_B''', R_{M-O}''') should be assigned to oxygen atoms of hydroxyl groups adsorbed on the sample surface.

The comparison of the values for the difference between the binding energies of inner and outer electrons for HTSC samples and source reagents makes it possible to understand qualitatively the density distribution of electron states in the valency band (∞ 0-10 eV) [7] (see fig. 10 and the table). Therefore, the top of this band can be estimated to consist of the states including the Tl $6s^2$, Tl 6p, Ba $6s^2$ and O $2p^6$ electrons. The states resulting mainly from the Cu 3d electrons are in the middle of the band at 3.5 eV. For more rigorous interpretation it is necessary to use the results of calculations.

The spectrum of low-energy electrons (∞10-40 eV) for HTSC samples has a more complex structure than that obtained by a simple superposition of spectra from atoms contained in the sample (fig. 10).

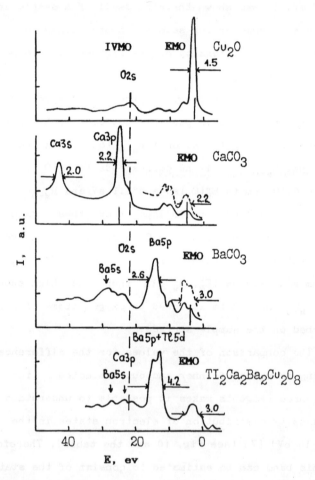

Fig. 10. XPS spectra of low-energy shell electrons for Cu_2O, $CaCO_3$, $BaCO_3$ and for sample with T_c = 103 K. The linewidth is reduced to $\Gamma(C\ 1s)$ = 1.3 eV.

This can be explained by a strong interaction of the Tl 5d, Ca 3p, Ba 5p, Ba 5s and O 2s shells of the neighbouring atoms to yield inner valent molecular orbitals (IVMO). It was shown for a great number of compounds of a variety of elements [8] that the IVMO formation leads to attenuation of the binding energy for electrons of some outer valent MO. The study of the contribution from the low-energy shell of atoms contained in HTSC samples to the MO formation can promote the understanding of the mechanism of forming the conditions for superconductivity.

The Tl-Ca-Ba-Cu-O ceramics considered has a great deal in common in its structure, including the electronic one, with the HTSC ceramics $Y_1Ba_2Cu_3O_{9-x}$ [5] studied earlier. It is possible to conclude on the basis of XPS data that the chemical binding in the ceramics studied is largely of a "metallic" nature.

The data obtained in the present work on the degree of oxidation of ions in the Tl-Ca-Ba-Cu-O metal-oxide ceramics indicate that the effective size of ions of thallium, calcium, barium and copper is essentially larger than that of Tl(3), Ca(2), Ba(2) and Cu(2) the radius of which are 0.105 nm, 0.104 nm, 0.138 nm and 0.080 nm, respectively [9]. The radius of mono-valent ions Tl(1), Cu(1) and O^{2-} amount to 0.136 nm, 0.098 nm and 0.136 nm, respectively. The appearance of weakly-oxidized ions in the HTSC ceramics under consideration allows one to explain the relatively large parameters of tetragonal lattice a=b=0.5453 nm and to construct the scheme of its atomic arrangement (fig. 11). On the basis of this scheme it is possible to assume that in the superconductive Tl-Ca-Ba-Cu-O ceramics there exist mainly two phases: 2122 and 2223 the separation of which is hindered due to a close resemblance between their structures differing, as seen

in fig. 11, in principle from the model of structure proposed in
[10].

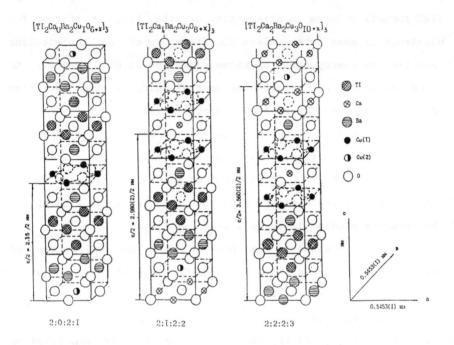

Fig. 11. Schematic structure of metaloxide ceramics Tl-Ca-Ba-Cu-O reconstructed with XPA and XPS data taken into account.

References

1. Sheng Z.Z., Hermann A.M., Nature, 1988, v. 332, N 6159, p. 55-58.
2. Sheng Z.Z., Hermann A.M., Nature, 1988, v. 332, N 6160, p. 138.
3. Hermann A.M., Sheng Z.Z., Abstract of YCMC, June 7-10, 1988, Shenyang. China, DD2.
4. Taldenkov A.N., Shabanov S.Yu., Inyushkin A.K., Babushkina N.A., Kopylov A.V., Florent'ev V.V. - In book: Superconductivity, Ed. by Ozhogin V.I. - Moscow, IAE, 1987, pp. 11-14 (in Russian).
5. Teterin Yu.A., Baev A.S., Sosul'nikov M.I., Simirsky Yu.N., Superconductivity. Ed. by Ozhogin V.I. - Moscow, IAE, 1988, issue 3, pp. 24-33 (in Russian).
6. Nefedov V.I., Gati D., Dzhurinsky B.F., Sergushin N.P., Salyn' Ya.V . Zh. neorg. khimii, 1975, v. 20, N 9, pp. 2307-2314 (in Russian).
7. Bondarenko T.N., Teterin Yu.A., Baev A.S. DAN SSSR, 1984, v. 279, N 1, pp. 109-113 (in Russian).
8. Teterin Yu.A., Gagarin S.G. - Moscow, TsNIIAtominform, 1985, (in Russian).
9. Brief chemical encyclopaedia. - Moscow, Soviet Encyclopaedia Publ., 1963, v. 2, p. 310 (in Russian).
10. Glosh P., New Sci, 1988, v. 118, N 1607, p. 28.

STRUCTURE GENESIS AND THE PHYSICAL PROPERTIES
OF PEROVSKITE-LIKE PHASES

N.E.Alekseevskii[1], D.Wlosewicz[1], A.V.Mitin[1], V.I.Nizhankovskii[1],
V.N.Narozhnyi[2], E.P.Khlybov[2], G.M.Kuzmicheva[3],
I.A.Garifullin[4], N.N.Garifianov[4], B.I.Kochelaev[5],
A.V.Gusev[6], G.G.Deviatykh[6], A.V.Kabanov[6]

[1] Institute for Physical Problems, Moscow
[2] Institute of High Pressures, Troitsk
[3] Institute of Fine Chemical Technology, Moscow
[4] Phisical-Technical Institute, Kazan'
[5] Kazan' State University, Kazan'
[6] Institute of Chemistry, Gor'kii

Quasi-layered structures are produced in metal-oxide systems among which the phases that pass into the superconducting state at high temperatures are found [1, 2]. These structures consist of Cu-O layers with the atoms of other elements such as Ba or rare earths situated between them. This paper gives an account of the crystal-chemistry structural analysis of some high-temperature superconductors and presents the results of the studies of superconducting and magnetic properties including the studies at high hydrostatic pressures.

Structural Features

As is known, there exists a small quantity of basic structures that may lead over to all other structures by way of ordering /superstructure formation/, insertion /subtraction/, or deformation.

The structure of cubic perovskite with ABO_3 as general formula is a basic one in the metal oxide systems in which the high-temperature superconducting phases are found /Fig. 1a/.

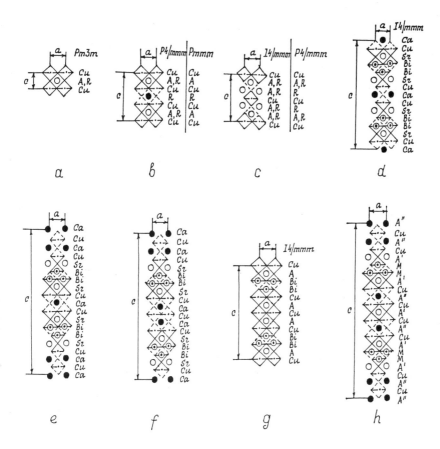

Fig. 1. Superconducting phase genesis with perovskite-like structure

The $SrTiO_{3-\delta}$ ($T_c \approx 0,43K$) and $(Ba_{0,6}K_{0,4})BiO_{3-\delta}$ ($T_c \approx 29,8K$) superconducting phases are crystalized just in the cubic perovskite structure.

The perovskite cubic structure considered polyhedrically is based on octahedra (BO_6). In this case the phase cell parameters with a perovskite structure may be estimated as $a = 2(r_B+r_0) = a_0$, $c = a_0$, where r_B is the B cation radius /radius of copper in metal oxide systems/, r_0 is the anion radius /here oxygen/.

In the R-Ba-Cu-O system /R is rare earths or Y/ there are produced four phases with $RBa_2Cu_3O_{6+\delta}$ as general formula which are the superstructure of the cubic perovskite /Fig. 1b/. These phases are crystallized in rhombic and tetragonal systems and differ from each other in the elementary cell parameter ratio: "ortho-1" phase (b>a, c = 3b), "ortho - II" phase (b>a, c>3b), "tetra-1" phase (b = a, c = 3b), "tetra-II" phase (b = a, c>3B) [3]. The rhombic phases are superconducting ones and the $T_c \approx 95K$ of the "ortho-I" phase is larger than the $T_c \approx 60K$ of the "ortho-II phase. In terms of polyhedra (CuO_6) the elementary cell parameters of these phases are $a=a_0$, $c=3a_0$ (Fig. 1b).

The samples of $(R, Ba)_2CuO_{4-\delta}$ and $(R, Sr)_2CuO_{4-\delta}$ are crystallized in K_2NiF_4 structure type ($T_c \approx 40K$). The structural type may also be considered as a derivative of the perovskite structure. In idealized form the phase cell parameters with the K_2NiF_4 type structure may be estimated in terms of octahedron packing (CuO_6) as $a=a_0$, $c=(2+\sqrt{2})\cdot a_0$ (Fig. 1c). Two phases with either body-centered or primitive cells can presumably be produced depending on the type of crystallographic ordering of the R and Ba (Sr)(Fig. 1c). The type of ordering depends both on composition and synthesis conditions and may vary depending on this or that way of action on the sample, which leads to a structural phase transition.

On the curves $\tilde{\chi}(T)$ of the two-phase samples with the $RBa_2Cu_3O_{6+\delta}$ and K_2NiF_4 type structures there are two fractures due to transitions of these phases to the superconducting state at $T_c \approx 90K$ and $T_c \approx 40K$ respectively.

The studies in the Bi-Ca-Sr-Cu-O system have demonstrated that three superconducting phases may be singled out with cell parameters: a=3.831, b=3.878, c=11.63Å (I); a=3.816, c=3050Å (II) and a=3.79, c=36.8Å (III). Phase I is similar to the $RBa_2Cu_3O_{6+\delta}$ phase of the R-Ba-Cu-O system (Fig. 1b). On the curve $\tilde{\chi}(T)$, however, there is only one fracture corresponding to $T_c \approx 90K$. Phase II is identical to the $Bi_4(Ca,Sr)_6Cu_4O_{16+\delta}$ phase [4]. Cell parameters of this phase can be estimated in terms of polyhedra (CuO_6) as follows: $a=a_0$, $c=(6+\sqrt{2})\cdot a_0$ (Fig. 1d). T_c of this phase is appr. 90K.[I)] Phase III of the $Bi_2Ca_2Sr_2Cu_3O_{10+\delta}$ composition with $T_c \approx 110-120K$ may have a structure presented in Fig. 1e or f [5], the cell parameters being $a=a_0$, $c=(8+\sqrt{2})a_0$. In the same system we have obtained a non-superconducting phase with cell parameters: a=3.805, c=24.51Å ($a=a_0$, $c=(5+\sqrt{2})a_0$). Fig. 1g presents one of the possible type of structure of this phase.

[I)]For some samples it is observed curves $\tilde{\chi}(T)$ and $\rho(T)$ with two fractures at appr. 50 and 90K (Fig.2). The fracture at 90K is connected with phase II and one at 50K - with phase IV (a=3.77, c=24.1Å).

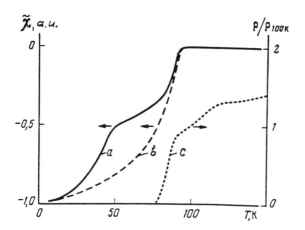

Fig. 2. Superconducting transitions of three Bi-Ca-Sr-Cu-O samples measured by their susceptibility in a.c. current (a,b) and by their resistance (c)

In the Tl-Ca-Ba-Cu-O system we have obtained two superconducting phases, the one of $Tl_2CaBa_2Cu_2O_{8+\delta}$ composition with pseudotetragonal cell parameters: a=3.84, c=29.10Å $(a=a_0, c=(6+\sqrt{2})a_0)$, $T_c \simeq 90-100K$ and the other of $Tl_2Ca_2Ba_2Cu_3O_{10+\delta}$ composition with cell parameters: a=3.82, c=36.2Å $(a=a_0, c(8+\sqrt{2})a_0, T_c \simeq 120K$. The X-ray data for these phases agree with those reported in [6] /we present the subcell parameters/. Similarity in the structure of these two phases in the Bi-Ca-Cr-Cu-O system is not ruled out (Fig. 1d,e or f). The superconducting phase with cell parameters a=3.84, b=3.86, c=11.57Å $(a=a_0, c=3a_0,$ Fig. 1b) and $T_c \simeq 90K$ is found in the Tl-Ba-Cu-O system, while the non-superconducting phase with a=3.87, c=23,24Å $(a=a_0, c=(5+\sqrt{2})a_0)$ is found in the Tl-Be-Ba-Cu-O system.

These arguments may offer a good reason for the prediction of a phase of the $M_4A_{10}Cu_8O_\delta$ ($M_4A_5'A_5''Cu_8O_\delta$) composition (Fig. 1h) with cell pseudotetragonal parameters $a=a_0$, $c=(10+\sqrt{2})a_0$) and $T_c > 120K$ since the intensity of low-frequency modes in the structures of this type may go up with c parameter growing, which must presumably lead to an increase in the T_c. But in this case, however, one must expect a decrease in lattice stability. It is quite possible that the reports on superconductivity at $T_c > 150K$ relate to these non-stable phases, and poor reproduceability and degradation of their properties in the cours of time just may be due to this phenomenon.

As follows from the consideration of the structure genesis [7], complication of the elementary cell of a superconducting phase brings about a distinctive "polymerization" of the initial cubic perovskite just as it was the case with molybdenum superconducting chalcogenides [8].

Magnetic Properties

Magnetic susceptibility was measured by the balance using Faraday's method [9]. In Fig. 3 there are shown the value of $\chi^{-1}(T)$

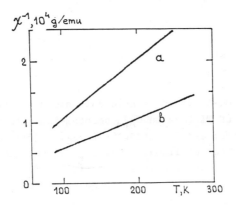

Fig. 3. Temperature dependence of inverse magnetic susceptibility for (a) $GdBa_2Cu_3O_{6+\delta}$ and (b) $DyBa_2Cu_3O_{6+\delta}$

for $RBa_2Cu_3O_{6+\delta}$ samples that incorporate rare earths with localized magnetic moments. These dependences are well described by Curie-Weiss' law where μ_{eff} values are close to those calculated in assumption of free ions R^{3+} (Table 1) and the negative sign of Curie's temperature Θ points to the process of antiferromagnetic ordering. Such ordering was actually observed in $GdBa_2Cu_3O_{6+\delta}$ [10].

Table 1. Magnetic characteristics of samples that have R ions with localized magnetic moments. The effective moment values are presented on the one R ion basis.

Composition	μ_{eff}, μ_B	Θ, k	$\mu_{R^{3+}}, \mu_B$
$NdBa_2Cu_3O_{6+\delta}$	3,5	−20	3,6
$GdBa_2Cu_3O_{6+\delta}$	7,7	−5	7,9
$DyBa_2Cu_3O_{6+\delta}$	11,1	−16	10,6
$TmBa_2Cu_3O_{6+\delta}$	7,8	−16,5	7,6
$Ho_{1,2}Ba_{0,8}Cu''_{4-\delta}$	9,9	−14	10,6
$Er_{1,2}Ba_{0,8}CuO_{4-\delta}$	8,7	−14	9,6
$Yb_{0,98}Ba_{0,72}CuO_{4-\delta}$	4,3	−40	4,5

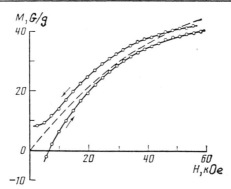

Fig. 4. Magnetic field dependence of $GdBa_2Cu_3O_{6+\delta}$ magnetic moment at T=4.2K.

In Fig. 4 is shown a dependence of the $GdBa_2Cu_3O_{6+\delta}$ magnetic moment on the field measured by a string magnetometer at T=4.2K. As the magnetic field is increased within 6 KO_e the field moment keeps negative as it must be in the case of ordinary superconductors. But in the fields over 6 KOe the magnetic field becomes positive and it shows a trend to saturation as the magnetic field is further increased. In this case the sample continues to be in a superconducting state which is indicated by a hysteresis of the magnetic moment. It should be mentioned that an averaged dependence of the magnetic moment on the field is fairly well described by Brillouin's function with S=7/2 which corresponds to a full moment of the free ion Gd^{3+} /dash line in Fig. 4/, and small departures from it may be due to proximity of the experimental temperature to the point of antiferromagnetic ordering T_N=2.2K [10].

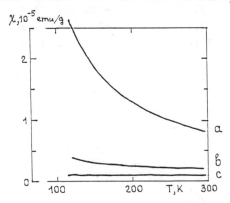

Fig. 5. Temperature dependence of magnetic susceptibility of
(a) $Tl_2CaBa_2Cu_2O_{8+\delta}$; (b, c) $TlCa_2Ba_2Cu_3O_{10+\delta}$

The magnetic susceptibilities of Y-Ba-Cu-O and Tl-Ca-Ba-Cu-O strongly depended on the phase composition and the way of preparation of samples. Some dependences of $\chi(T)$ are given in Fig. 5. These dependences are well described by the equation $\chi = \chi_0 + \chi_{cw}$, where χ_0 is independent of temperature and χ_{cw} is described by Curie-Weiss' law. It has been noticed that the value of χ_{cw} is decreased with an increase in the superconducting phase. The value of χ_0 independent of temperature amounts to 7.5×10^{-7} emu/g for $YBa_2Cu_3O_{6+\delta}$ and to 8.0×10^{-7} emu/g for $Tl_2Cu_2Ba_2Cu_3O_{10+\delta}$. Assuming the value of χ_0 dependent on the Pauli spin paramagnetism, the estimated electronic heat capacity γ will be $\gamma \approx 37$ mJ/mol·K^2 for $YBa_2Cu_3O_{6+\delta}$ and $\gamma \approx 65$ mJ/mol·K^2 for $Tl_2Ca_2Ba_2Cu_3O_{10+\delta}$.

A detailed information on the nature of paramagnetic centers and on the microstructure of their surrounding can be obtained through the EPR in $YBa_2Cu_3O_{6+\delta}$. The EPR measurements were made by the BER-418[S] spectrometer with a frequency of 9400 MHz within 1.5K and 300 K [11]. Fig. 6a shows an EPR spectrum typical for $YBa_2Cu_3O_{6+\delta}$ with temperatures above the T_c. Its appearance is specific for powdered samples that incorporate ions with an unisotropic g-factor of axial symmetry. The numerical simulation has demonstrated that $g_{par.}=2.200\pm0,005$ and $g_{norm.}=2.060\pm0.005$. These values are characteristic of the EPR signals of ions Cu^{2+} in an axial crystalline field with a basic state described primarily by the wave function $d_{x^2-y^2}$ and not by that of d_{z^2} for which $g_{norm.} > 2.0023 > g_{par.}$ might be expected [12]. Rough estimates demonstrate that appr. 10 per cent of copper ions in the $YBa_2Cu_3O_{6+\delta}$ phase take part in resonance. That makes it possible to attribute an χ_{cw} temperature-dependent part of magnetic susceptibility to the presence of ions Cu^{2+}.

To clarify the arrangement of localized moments in $YBa_2Cu_3O_{6+\delta}$ the EPR values were measured on oriented samples. For this purpose the powder particles of individual crystalline grain size $(2-7\mu)$ was mixed with paraffin. Than the ampule with the mixture was placed in the magnetic field of H=10-50 kOe. Paraffin was heated to the temperature of melting and then cooled in magnetic field. An χ-ray analysis has demonstrated that due to this procedure, the

crystallographic axis C of a significant number of cristalline grains becomes oriented along the magnetic field. The EPR spectra based on two orientations of the sample with respect to magnetic field $\vec{H}$ are shown in Fig. 6c, e.

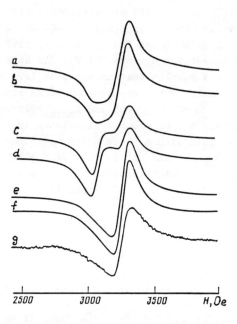

Fig. 6. EPR spectra in $YBa_2Cu_3O_{6+\delta}$ at T=130K: (a) powder sample; (c) oriented sample with $\underline{c}$ parallel to $\vec{H}$; (e) oriented sample with $\underline{c}$ normal to $\vec{H}$; (b, d, f) numerical simulation results for $g_{par.} = 2.2$, $g_{norm.} = 2.06$, homogeneous line width is 100 Oe, oriented part makes up 60 percent.

An inference can follow that the axis of symmetry of local surrounding of ions Cu^{2+} coinsides with crystallographic axis $\underline{c}$. This inference is corroborated by numerical simulation (Fig. 6d, f).

It is most probable that paramagnetic centers of Cu^{2+} are between the barium ions in the planes incorporating both the copper and oxigen ion chains. Octahedric surrounding of ions Cu^{2+} corresponding to the spectrum $d_{x^2-y^2}$ is set up when vacancies between the chains are filled with oxigen.

The temperature dependence of the width and attitude of the EPR line is shown in Fig. 7.

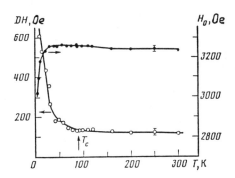

Fig. 7. Temperature dependence of resonant field and line width of the Cu^{2+} EPR in oriented $YBa_2Cu_3O_{6-\delta}$ with $\underline{c}$ normal to $\overrightarrow{H}$.

The absence of temperature-dependent linear increase in the Korringa contribution to the line width at $T > T_c$ indicates that the localized moments of Cu^{2+} are not in contact with band electrons.

It should be noted that, in accordance with the crystallochemical notion, the copper ions which are situated in the planes Cu-O most proximate to Y make up more than 2/3 of the total number of copper ions in $YBa_2Cu_3O_{6+\delta}$ must keep in bivalent state. The absence of temperature-dependent magnetic susceptibility in single phase superconducting samples and of a corresponding EPR signal may testify to the fact that in Cu-O layers most proximate to Y, ions Cu^{2+} become ordered antiferromagnetically with $T_N > 300K$ or form a collective band [13, 14].

In $GdBa_2Cu_3O_{6+\delta}$ we observed a single smooth EPR line with g=2.03±0.02 produced by ions Gd^{3+}. The observed linear temperature dependence of the line width DH with $\partial(DH)/\partial T = 0.82$ Oe/K (Fig. 8) is due to the Korringa relaxation of gadolinium moments attached to conduction electrons (Fig. 8).

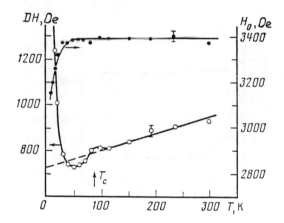

Fig. 8. Temperature dependence of resonant field and line width of the Gd^{3+} EPR in $GdBa_2Cu_3O_{6+\delta}$

$$DH_K = \frac{\pi k_B}{g \mu_B} (\rho_F J_{sf})^2 \cdot T \qquad (1)$$

where ρ_F is density of states at the Fermi level, J_{sf} is an integral of exchange interaction of $\underset{\sim}{f}$ electrons with conduction electrons.

Estimating ρ_F both from $\partial H_{c2}/\partial T$ and residual resistance one will have $\rho_F \approx 0.8$ $eV^{-1} atom^{-1} spin^{-1}$ hence $J_{sf} \approx 7.4 \times 10^{-3} eV$. Previously the small quantities of J_{sf} were observed in the Chevrel phases for rear earth ions [15]. Using J_{sf} and ρ_F one can estimate a decrease in the T_c due to localized magnetic moments.

$$\Delta T_c = -\frac{\pi^2}{8 k_B} S(S+1) \rho_F J_{sf}^2 \underset{\sim}{x} \qquad (2)$$

where S is spin of localized moments, $\underset{\sim}{x}$ is their concentration. For $GdBa_2Cu_3O_{6+\delta}$ the estimate will be $\Delta T_c = -0.75 K$. It well agrees with an experimentally slight dependence of the T_c on the sort of rare-earth ions in these systems [9].

A residual temperature-independent line width that may be obtained by linear extrapolation of DH(T) to 0 K is 720 Oe. It provieds a notion that the residual line width of a powder sample is determined by dipole-dipole interaction and by an unresolved fine structure. A full width of the fine structure of 3000 Oe was determined on the oriented $Y_{0.98}Gd_{0.02}Ba_2Cu_3O_{6+\delta}$. Dipole contribution to the line width amounted to 2500 Oe. An observed residual line width of 720 Oe is smaller than other allreadymentioned contributions to DH which is an indication of an exchange-induced narrowing of the EPR line. To estimate the exchange integral let us use an equation for the EPR line width used in the moments method.

$$DH = \pi(M_2^{3/2} M_4^{1/2})/2\sqrt{3} \qquad (3)$$

Taking into account the exchange interactions between the nearest neighbors and two coordination spheres for dipole interactions yields

$$M_2 = 2.5S(S+1)(g\mu_B)^4 a^{-6} + 15D^2$$
$$M_4 = 7.7S(S+1)(g\mu_B)^4 a^{-6} J^2 \, (Gd-Gd) \qquad (4)$$

Using $S=7/2$, $a \approx 3.8$ Å and measured $DH(0) = 720$ Oe and $D \approx 0.02$ cm^{-1} from (4) and (5) we shell have $J(Gd-Gd) \approx 0.12$K. In this case a Curie paramagnetic temperature $\theta = S(S+1) \cdot J(Gd-Gd)$ for the nearest neighbors $z=4$ amounts to 2.4K which is comparable with $\theta = -5$K obtained from the data on temperature-dependent magnetic susceptibility /Table 1/.

The exchange integral J_{sf} obtained from (2) permits to estimate an integral of the RKKI interaction between magnetic ions Gd^{3+} through conduction electrons. The estimation of this kind in approximation of free electrons yields an exchange integral $J(Gd-Gd)$ whose value is an order of magnitude less than that of the one obtained above. It is apparently an indication of an important role that the superexchange mechanisms and details of conductivity band structure play in exchange interaction between ions Gd^{3+}.

Transition of the rare-earth compound GdBa$_2$Cu$_3$O$_{6-\delta}$ to an antiferromagnetically ordered state at $T_N = 2.23$K was observed on measuring its heat capacity [10, 16]. The heat capacity [16] and magnetic susceptibility data [9] of GdBa$_2$Cu$_3$O$_{6-\delta}$ are readily interpreted as transition of the sublattice of ions Gd^{3+} with a spin of 7/2 into the ferromagnetic state in almost over the whole volume of the sample. With $T < T_N$ this compound remains to be in a superconducting state. Thus, in the case of this compound, there is observed a coexistence of superconductivity and magnetism, similar to that mentioned previously, say, for Chevrel rare earth phases.

We measured a heat capacity of GdBa$_2$Cu$_3$O$_{6+\delta}$ under pressure. Samples were measured at $2 \leq T \leq 4$ K with the aid of quasiadiabatic, calorimeter into which a myniature high-pressure cell made of nonmagnetic berillium bronze was placed. The pressure in the cell was evaluated by measuring the shift of the T_c of a tin manometer placed beside the sample in the cell channel. The channel diameter is 4 mm, cell's weight is about 50g, and sample's weight is appr. 0.3g. In the course of experiment we measured heat capacity of the cell with a sample under pressure and without pressure and heat capacity of the empty cell. To determine the heat capacity of a sample the empty cell heat capacity value was subtracted from the sample-filled chamber heat capacity value. As is seen from Fig. 9

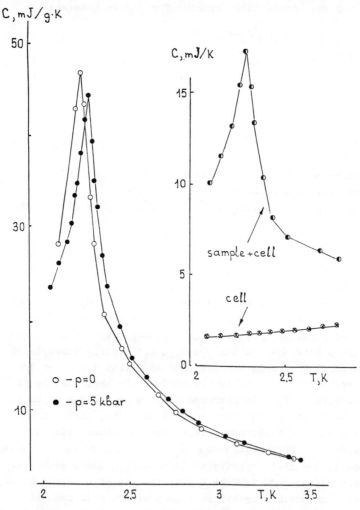

Fig. 9. $GdBa_2Cu_3O_{6+\delta}$ heat capacity measurement results in high-pressure cell.

a broadening of the peak of the sample under pressure did not occur. Absolute heat capacity of the sample at $T=T_N$ is an order of magnitude higher than that of the cell. It has provided confidence in singling out heat capacity of a relatively small sample close to the T_N. A shift of the T_N from 2.230K to 2.265K i.e. by 0,035K under the pressure of about 5 kbar was observed. The value of $T_N/\partial P$ was 7×10^{-6} K/bar.

At the present time there are different opinions about the nature of magnetic ordering in this compound. The results of measurement of the EPR on Gd^{3+} reported here can be interpreted in favor of the exchange character of interaction that leads to an antiferromagnetic transition in the compound. Comparison of the values of the T_N conpounds with those of different rare ea ths also makes the authors of paper [17] think about the exchange character of ordering. At the same time the paper [18] postulats that an antiferromagnetic ordering in $GdBa_2Cu_3O_{6+\delta}$ is due to a dipole-dipole interaction.

Estimation of the derivative $|\partial \ln T_N / \partial \ln V| = |æ^{-1} \partial \ln T_N / \partial P|$ using the compressibility value $æ = 0.87 \times 10^{-6}$ bar^{-1} [19] will give ~ 4.

It should be noted that for a dipole-dipole interaction the value of this derivative must be close to 1, because in this case $T_N \sim \mu^2_{eff}/d^3 \sim \mu^2_{eff} N$ [20]. Thus, the results of measurement of the heat capacity of $GdBa_2Cu_3O_{6+\delta}$ under pressure suggest a notion that the interaction is of exchange character that leads to an antiferromagnetic ordering in the compound.

Superconducting properties

We have measured heat capacities of some high-temperature superconductors in the region of transition to a superconducting state. Measurements were made by an adiabatic calorimeter. The mass of samples was about 3g. In order to prevent oxygen losses, samples were placed in a thin-walled leek-proof copper capsule filled with gaseous helium. The results are listed in Fig. 10. This figure

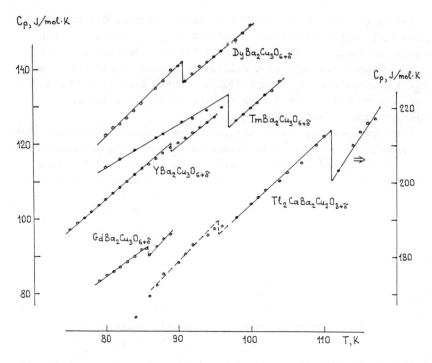

Fig. 10. Temperature dependence of heat capacity of a number of high-temperature superconductors close to the T_c.

shows that, although the transition to a superconducting state was slightly "smeared", the extrapolation of the Cp temperature dependences at $T > T_c$ and $T < T_c$ to $T = T_c$ permits to mark confidently a jump of heat capacity ΔCp at $T = T_c$. Data for the T_c, ΔCp and γ_c calculated by equation $\gamma_c = \Delta Cp / 1.43 \cdot T_c$ and for $\gamma\gamma$ determined from temperature-dependent contribution to magnetic susceptibility γ_0 are listed in Table 2.

Table 2. Heat capacity jumps (ΔCp) and electron heat capacity coefficients (γ)

Compound	T_c, K	$\Delta Cp, \dfrac{J}{mol \cdot K}$	$\gamma_c, \dfrac{mJ}{mol \cdot K^2}$	$\gamma\gamma, \dfrac{mJ}{mol \cdot K^2}$
$YBa_2Cu_3O_{6+\delta}$	89	2	16	37
$DyBa_2Cu_3O_{6+\delta}$	90,7	5,5	42	–
$TmBa_2Cu_3O_{6+\delta}$	97	8,6	62	–
$GdBa_2Cu_3O_{6+\delta}$	86	3,1	25	20
$Tl_2CaBa_2Cu_2O_{8+\delta}$	111	13,1	82	65

Table 2 shows a reasonable agreement of coefficients of electron heat capacities γ_c and γ_χ.

The second critical field H_{c2} was determined from resistive curves of superconducting transition $\rho(T,H)$ obtained either in stationary (to 150 kOe) or pulsed (to 500 kOe) megnetic fields [9]. The H_{c2} was taken as the field corresponding to a midpoint of the superconducting transition curve $(0,5\rho n)$. Fig. 11 shows a $H_{c2}(T)$

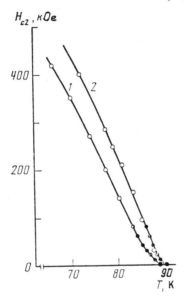

Fig. 11. Temperature dependence of the H_{c2} for Lu-Ba-Cu-O (1) and Y-Ba-Cu-O (2). (●) stationary field measurements, (O) pulsed field measurements.

relation for Y and Ln. The maximum values of derivative $|\partial H_{c2}/\partial T|_{max}$ were 25 kOe/K for Y and 20 kOe/K for Lu. If we use a ratio $H_{c2}(0) \approx 0.7 \cdot T_c \cdot |\partial H_{c2}/\partial T|_{max}$ to estimate $H_{c2}(0)$ than for Y-Ba-Cu-O we shall have $H_{c2}(0) \approx 1.5$ MOe. Estimating the coherence length $\xi = \sqrt{\Phi_0/2\pi H_{c2}(0)}$ in this case will give appr. 8 Å which exceeds the value of the lattice parameter $\underset{\sim}{a}$ only by a factor of two.

The critical fields H_{c2} for Bi-Ca-Cr-Cu-O are somewhat lower than those for R-Ba-Cu-O; the value of $|\partial H_{c2}/\partial T|_{Max}$ for the sample of a second structure type (a=3,87 Å, C=30,62 Å) with T_c=88K was 11 kOe/K.

For $Tl_2Ca_2Ba_2Cu_3O_{10-\delta}$ the T_c taken against the level of $0.9\rho n$ was 124K and $|\partial H_{c2}/\partial T|_{Max}$ amounted to 25kOe/K.

It should be emphasized in conclusion that the results reported in this paper for three related perovskite-like metal oxide phases of R-Ba-Cu-O, Bi-Ca-Sr-Cu-O, and Tl-Ca-Ba-Cu-O are somewhat alike and, in principle, do not differ from the results obtained for ordinary second type superconductors. A week influence of rare earths on the superconducting properties is due to a specific crystall structure of these compounds.

This paper presents the studies of systems with two layers of Cu-O in the elementary cell. As has already been mentioned previously in [7], in systems with a larger number of layers one can expect an increase in the T_c above 120 K. Recent communication [21] has reported data about the superconductivity at T_c=162K in Tl-Ca-Ba-Cu-O with three layers of Cu-O in the elementary cell. The structure of this compound apparently corresponds to that shown in Fig. 1h. As it was expected [7] this compound proved very unstable [21].

References

1. J.G.Bednorz, K.A.Müller: Z.Phys. B64, 189 (1986)
2. M.K.Wu, J.R.Ashburn, C.J.Torng, P.H.Hor, R.L.Mehg, L.Gao, Z.J.Huang, Y.Q.Wang, C.W.Chu: Phys.Rev.Lett. 58, 908 (1987)
3. M.Ishikawa, T.Takabatake, Y.Nakazawa: Technical Report of ISSP A 1908, 30 (1988)
4. J.M.Tarascon, Y. Le Page, P.Barboux, B.G.Bagley, L.H.Greene, W.R.Mc Kinnon, G.W.Hull, M.Giroud, D.M.Hwang: Phys.Rev. B37, 2 (1988)
5. J.M.Tarascon, W.R.Mc Kinnon, P.Barboux, D.M.Hwang, B.G.Bagley, L.H.Greene, G.Hull, Y. Le Page, N.Stoffel, M.Giroud: preprint
6. R.M.Hazen, L.W.Finger, R.J.Angel, C.T.Prewitt, N.L.Ross, C.G.Hadidiacos, P.J.Heaney, D.Veblen, Z.Z.Sheng, A.El Ali, A.M.Hermann: Phys.Rev.Lett. 60, 1657 (1988)
7. N.E.Alekseevskii, G.M.Kuzmicheva, E.P.Khlybov, A.V.Mitin, V.I.Nizhankovskii: Pis'ma Zh. Eksp. Teor. Fiz. 48, 45 (1988)
8. M.Potel, R.Chevrel, M.Sergent: Acta Crystall. B56, 1319 (1980)
9. N.E.Alekseevskii, E.P.Khlybov, G.M.Kuzmicheva, V.V.Evdokimova, A.V.Mitin, V.I.Nizhankovskii, A.I.Kharkovskii: Zh. Eksp. Teor. Fiz. 94, no. 5, 281 (1988)
10. J.O.Willis, Z.Fisk, J.D.Thompson, S.W.Cheong, R.M.Aikin, J.L.Smith, E.Zirngiebl: J.Magn.Magn.Mat. 67, L139 (1987)
11. N.E.Alekseevskii, I.A.Garifullin, N.N.Garifianov, B.I.Kochelayev, A.V.Mitin, V.I.Nizhankovskii, L.P.Tagirov, E.P.Khlybov: Zh. Eksp. Teor. Fiz. 94, no. 4, 276 (1988)
12. A.Abragam, B.Bleaney 'Electron Paramagnetic Resonance of Transition Ions', Clarendon Press, Oxford, (1970)
13. J.Rossat-Mignod, P.Burlet, M.J.G.M.Jurgens, J.M.Hentry,C.Vettier: Physica C152, 19 (1988)
14. J.H.Brewer, E.J.Ansalano, J.F.Cardan et al : Phys.Rev.Lett. 60, 1073 (1988)
15. R.Odermatt: Helv.Phys.Acta. 54, I (1981)
16. N.E.Alekseevskii, A.V.Gusev, G.G.Deviatykh, A.V.Kabanov, A.V.Mitin, V.I.Nizhankovskii, E.P.Khlybov: Pis'ma Zh. Eksp. Teor. Fiz. 47, 139 (1988)
17. A.P.Ramirey, L.F.Schneemeyer, J.V.Wasgesak: Phys.Rev. B36, 7145 (1987)

18. S.Harabhusan: Phys. Lett. A129, 131 (1988)
19. N.V.Aleksandrov, A.F.Goncharov, S.M.Stishov: Pis'ma Zh. Eksp. Teor. Fiz. 47, 357 (1988)
20. N.E.Alekseevskii, V.N.Narozhnyi, A.P.Dodokin, Kh.S.Bagdasarov: Dokl. Akad.Nauk USSR 281, 1094 (1985)
21. R.S.Lin, P.T.Wu, J.M.Liang, L.J.Chen: High T_c Update 2, no 16 (1988)

VORTEX STRUCTURE OF HIGH-T_c SINGLE CRYSTALS

L.Ya.Vinnikov, I.V.Grigorieva, L.A.Gurevich, Yu.A.Ossipyan

Institute of Solid State Physics, USSR Academy of Sciences, Chernogolovka, Moscow district, 142432, USSR

Abstract

The high-resolution Bitter technique has been used to reveal the magnetic structure of single crystal samples of high-T_c superconductors $YBa_2Cu_3O_x$ and $HoBa_2Cu_3O_x$ at 4.2 K. The lattice of Abrikosov vortices has been observed in the (ab) plane of the crystals. The flux line lattice appeared to be distorted that testifies, in our opinion, to the anisotropy of the magnetic field penetration depth λ which is in turn related to the anisotropy of the carrier effective mass. Besides, the volume pinning force and critical currents are determined from the decoration patterns for crystals with different density of the twin boundaries.

A direct observation of the type-II superconductor magnetic structure makes it possible to study a rather wide set of problems related to the flux line lattice (FLL) structure and vortex distribution. These problems include quantization of the magnetic flux, bulk or local character of superconductivity, possible distortions of the regular FLL caused, for instance, by the anisotropy of the superconducting properties. Besides, it gives a possibility for assesment of such fundamental characteristics as the magnetic field penetration depth λ and the lower critical field H_{c_1}; for a study of pinning effects that

are decisive in the lossless current-carrying capability of
superconductors. Solving these problems has become especially
urgent due to the discovery of the new class of high-T_c superconductors /1/ that are superconductors of type-II, with large
GL parameter $æ$ and highly anisotropic properties.

First observations of the high-T_c superconductor magnetic
structure on single crystals using the high-resolution Bitter
pattern technique /2/ are presented in /3,4/. In the present
paper we are summarizing our results in this field that are
partly published already /5/.

Samples and experimental technique

Samples under investigation were $YBa_2Cu_3O_x$ and $HoBa_2Cu_3O_x$
single crystals grown by slowly cooling the melt of nonstoichiometric oxide composition /6/. The samples were flat platelets
$\simeq 1$ mm in size along $\bar{a}$- and $\bar{b}$-axes and $10\div100\,\mu m$ parallel to
the $\bar{c}$-axis. The crystal surfaces were optically smooth and not
treated additionally.

The magnetic structure was investigated by the decoration
technique using disperse Fe particles /7/ at 4.2 K. The samples
were cooled in the external magnetic field down to the liquid
helium temperature (so called frozen flux regime). After the
decoration the sample surfaces were examined immediately either
in a scanning, or in an optical microscope (if the external magnetic field was lower than 20 Oe). The obtained vortex patterns
were optically treated with a lazer diffractometer. Angles of
reflexes in the diffractograms were measured with a micrometer
microscope; the accuracy of the angle measurements was $\simeq 1°$
(with account taken of the reflex smearing).

Experimental results and their discussion

Our observations of the crystal surfaces in the optical
microscope in polarized light showed twin domains, as a rule,
of two perpendicular orientations along the $\{110\}$ type planes.
The dimensions of the domains optically resolved were from several fractions of micrometer to $100\,\mu m$ in different sections
of the crystals. Fe particles were evaporated when the external
magnetic field took two different values: $H_e=10\div15$ Oe and $H_e=$

100 Oe, and two orientations: $\bar{H}_e \perp \bar{c}$ and $\bar{H}_e \parallel \bar{c}$. Examining the decoration patterns has demonstrated the following.

1. The triangular flux line lattice shown in Fig.1 is observed in the $(\bar{a}\bar{b})$ plane for the crystal sections free of twins. This result, first of all, is an evidence for the bulk superconductivity in the investigated materials. Besides, it demonstrates that the magnetic field perpendicular to the $\bar{c}$-axis penetrates a high-T_c superconductor in form of flux quantums. A flux value per single vortex Φ_1 can be found from comparison of the vortex number per unit area n and the induction $B \simeq H_e$ (the induction was measured by means of the Hall gauge placed immediately under the specimen; the demagnetization factor $D \simeq 1$). The above comparison showed Φ_1 to be practically equal to $\Phi_0 = 2 \cdot 10^{-7}$ G·cm^2. Thus the lattice of Abrikosov vortices, each carrying a flux quantum $\Phi_0 = hc/2e$, is observed in the $(\bar{a}\bar{b})$ plane that testifies to the current transfer by pairs of carriers. Starting from the observed ferromagnetic particle

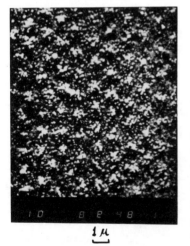

1 μ

Fig.1. The lattice of Abrikosov vortices in YBa$_2$Cu$_3$O$_x$ single crystal. Fe particle agglomerations present single vortices. (The photo is taken in the scanning electron microscope).

agglomeration dimensions in the decoration patterns one can estimate the penetration depth λ and, correspondingly, the lower critical field H_{c_1} /4/. Under optimal decorating conditions the ferromagnetic particles are congregated near the maximum magnetic field gradient around the vortex core. This region at the sample surface is about $2\lambda(T)$ in size. Despite some uncertainty in finding the "vortex diameter" which makes the λ estimate rather approximate, in the time being it may be of interest because $\lambda(0)$ values for YBa$_2$Cu$_3$O$_x$ found by different methods vary from 21 Å to 4500 Å /8/. We estimate $\lambda(4.2\,K) \lesssim 0.3\,\mu m$. Taking $æ = 100$

we obtain a lower estimate for H_{c_1}:

$$H_{c_1} = \frac{\Phi_o}{4\pi \cdot \lambda^2} \ln \varkappa > 100 \text{ Oe}$$

2. Effects associated with anisotropy of the superconducting properties of $YBa_2Cu_3O_x$ in the $(\bar{a}\bar{b})$ plane presented another important feature of the vortex structure in the free-of-twin regions of the crystals.

Analyzing the FLL images by optical diffraction /7/ we found stable and reproducible distortions of the regular and triangular FLL in the $(\bar{a}\bar{b})$ plane. Figs. 2a and 2b show the FLL images in the twin-free sections and corresponding diffractograms for two values of the external magnetic field: H = 10 Oe and H = 100 Oe. Both patterns were obtained on the same sample. In the magnetic field H_e = 10 Oe the FLL is seen to be split into misoriented blocks of about several vortex spacings in size. In the field H_e = 100 Oe this fine-scale splitting is not observed. The described FLL behaviour is consistent with the growing intervortex interaction and increasing FLL rigidity on increasing the magnetic field that leads to an increase of the short-range-order regions in the FLL /10,11/. Nevertheless, the degree of the FLL distortions appeared to be practically the

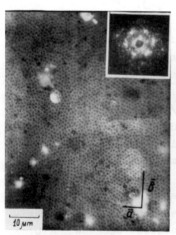

a

b

Fig.2. a) FLL in the twin-free section of the $YBa_2Cu_3O_x$ single crystal. The mean induction $\bar{B}$=9 G. The diffractogram taken from this section is given in the insert.
b) the same that in (a) but the induction

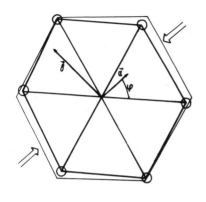

c

$\bar{B} \simeq 100$ G.

c) the scheme of the reflex shift in the diffractogram when compressing the regular triangular lattice along the $\bar{a}$-axis.

same in both cases. Some of the angles between the FLL triangular cell sides taken on the diffractograms for various crystal sections and magnetic fields $H_e \simeq 10$ Oe and $H_e \simeq 100$ Oe are given in Table 1. The observed deformation of the regular FLL can be definitively represented as its compression along the axis making an angle φ with one of the FLL vectors, with a compression factor η (see the scheme in Fig.2c). In this representation the compression axis appeared to be always the same for the same domain orientation (domains of the same orientation are identically coloured in polarized light) and perpendicular to the $\bar{b}$-axis found by the light phase delay using the Berek phase compensator /12/, i.e. the compression axis proved to be the $\bar{a}$-axis of the crystal. Values of the compression factor η were very close to each other for different crystal sections and practically independent of the magnetic field (in the least within the range from 10 to 100 Oe).(see Table 1). Its value $\eta = 1.2 \pm 0.1$ At the same time the angle φ could take practically all possible magnitudes that testified to the vortex row directions and crystal axes not to be rigidly linked with each other.

We relate the observed FLL distortions to the anisotropy of the $YBa_2Cu_3O_x$ superconducting properties in the $\bar{a}\bar{b}$ plane. According to the theoretical predictions, the anisotropy of superconducting carrier effective mass μ should lead to the compression of the FLL along the axis corresponding to larger μ /13,14/, with the compression factor $\eta = \mu_{\bar{a}}/\mu_{\bar{b}}$ $(\mu_{\bar{a}} > \mu_{\bar{b}})$ even in the case of widely-spaced FLL with $d \gg \lambda$ (d is the mean vortex spacing) /14/. The FLL self energy is degenerate over the ψ angle between the compression axis and FLL vectors.

Starting from the agreement between the theoretical predictions and experimentally observed FLL deformation we can estimate the ratio of the effective mass tensor components in $\bar{a}$ and $\bar{b}$ directions using the obtained compression factors for $YBa_2Cu_3O_x$ $\mu_{\bar{a}}/\mu_{\bar{b}} = 1.4 \pm 0.2$.

Table 1

H_e	α	β	γ	η	ψ
100 Oe	66°	60°	54°	1.16	17°
	67°	57°	55°	1.16	28°
	71°	55°	54°	1.20	35°
	67°	58°	55°	1.14	28°
10 Oe	67°	58°	55°	1.14	28°
	65°	63°	52°	1.16	6°
	69°	59°	52°	1.24	18°
	64°	62°	54°	1.12	6°

We also attempted to investigate the magnetic flux distribution when the external magnetic field was perpendicular to the $\bar{c}$-axis, i.e. at the ($\bar{a}\bar{c}$) and ($\bar{b}\bar{c}$) faces. The result appeared to be rather different from the described above. Namely, for $HoBa_2Cu_3O_x$ in the external field $H_e \simeq 10$ Oe parallel to the $\bar{c}$-axis we observed the vortex structure similar to that shown in Fig.1. But in the same field $H_e \simeq 10$ Oe but perpendicular to the $\bar{c}$-axis we failed to resolve the vortices and observed a uniform Fe particle distribution at the surface. The same patterns were obtained on lowering the external magnetic field $\bar{H}_e \perp \bar{c}$ down to $H_e \simeq 1$ Oe. The uniform distribution of Fe particles is believed to indicate the magnetic flux presence inside the specimen, but more concrete conclusions about its structure are unattainable.

The reason for our failure to resolve the vortex structure in ($\bar{a}\bar{c}$) and ($\bar{b}\bar{c}$) planes is likely to result from too large λ value along the $\bar{c}$-axis compared with those along $\bar{a}$- and $\bar{b}$-axes. A considerable difference between $\lambda(\bar{c})$ and $\lambda(\bar{a},\bar{b})$

should be expected, for instance, from measurements of the H_{c1} anisotropy which was found to be about 6 /14/. On increasing λ the magnetic field gradient near the vortex core drops rapidly and, correspondingly, the force $\mu \cdot \text{grad } \bar{H}$ on the Fe particles also drops. So the decrease of the field gradient is due to the increase of λ and simultaneous decrease of H_{c1}. These two factors result in declining the resolving power of the decoration technique.

3. Now let us consider the effects of vortex pinning which could be directly observed in the crystal sections containing twin boundaries. When the external magnetic field is H 20 Oe and twin boundary spacings exceed 0.5 μm the vortex - twin boundary interaction can be observed directly in an optical polarizing microscope. Fig.3a shows a characteristic surface section image taken with uncrossed polaroids. Single vortices and their distribution through the sample are distinctly resolved. Examining the same section with crossed polaroids when all the twins are visible very well, one can follow an obvious correlation between the vortex and twin boundary relative positions (see Fig.3b). Namely, the vortices are localized along the twin boundaries that is likely to indicate their mutual attraction.

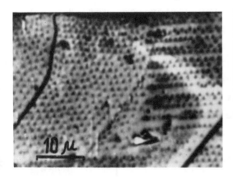
a

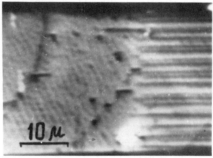
b

Fig.3. Two images of the same section of the $YBa_2Cu_3O_x$ single crystal taken in the polarizing optical microscope when a) polaroids are uncrossed (dark points are the vortices); b) polaroids are crossed (twins are visible).

The fact of vortex attraction to the twin boundaries conflicts with the representation of the latter as regions with a higher than in the matrix temperature of superconducting transition /15/ because it predicts a repulsion between the vortex and twins. On the contrary, the attraction indicates that the twin boundary is a "weak link", i.e. a region with lower than in the matrix transition temperature (a region with lower T_c is profitable for a normal vortex core to arrange in /16/). A very effective pinning ability of the twin boundaries in high-T_c superconductors is likely to be related to the small coherence length ξ compared to a boundary thickness. Our investigations of the vortex arrangement in the crystal sections containing nonuniformly distributed twin boundaries allowed to estimate quantitatively the volume pinning force F_p and critical current j_c in those samples.

Firstly, we shall find a pinning force of a single boundary. This appears possible due to the presence of large defect-free regions, where the FLL is retained, and single twins "embedded" in them (as shown in Fig.3). After that we shall estimate the pinning force and critical currents for crystals with different defect density.

As seen in Fig.3, the vortices pinned by the "embedded" twins are spaced exactly two times closer than those in the surrounding lattice. It seems as if two vortex rows were "inserted" one into another. A work of the pinning forces holding the vortices in the twin region is obviously equal to a work of the vortex repulsion forces. From this the force of the vortex-twin boundary interaction f_p can be estimated as the maximum repulsion force between two vortex rows. The desired force $f_p \geqslant F_r$. The F_r value as a derivative of the vortex interaction energy /18/ is given by the following expression:

$$F_r = (\frac{\Phi_o}{2\pi\lambda})^2 \left[(\frac{\pi \cdot \cos\alpha}{2\lambda \cdot r})^{1/2} + (\frac{\pi \cdot \lambda \cdot \cos^3\alpha}{8 r^3})^{1/2} \right] \exp(-\frac{r}{\lambda \cdot \cos\alpha}) \cdot \cos\alpha$$

where r is the distance from the vortex row corresponding to the maximum interaction; 2α is the angle for a vortex "to see" its neighbours in the opposite row. As a numerical calculation has shown the maximum F_r corresponds to $r = 0.18$ d. Substituting the numerical values $d = 1.5 \cdot 10^{-4}$ cm, $\lambda = 10^{-5}$ cm, we obtain the pinning force from a single twin boundary per unit area: $f_p \geqslant 0.5$ dyn/cm^2.

Going from the obtained individual pinning force it becomes possible to predict a volume pinning force F_p and critical current j_c if the twin density in the crystal is known. If the twin boundary spacing exceeds the vortex spacing d considerably then each boundary interaction with the FLL can be considered independently and $F_p = n \cdot f_p$, where n is the number of the twin boundaries per unit volume. In our samples the boundary spacings, as a rule, were smaller than or comparable with d in low magnetic fields and it was necessary to take account of simultaneous influence on the FLL from many defects. In this situation it is reasonable to use the idea of collective pinning /9/: $F_p = f_p \cdot (n/V_c)^{1/2}$, where V_c is the correlation volume. For the sample shown in Fig.4 the FLL is completely destroyed by pinning and V_c should be supposed as $V_c \simeq d^3$. Then, including the twin boundary spacing $\simeq 5 \cdot 10^{-5}$ cm we obtain $F_p \simeq 10^7$ dyn/cm^3.

Fig.4. A typical section of the $YBa_2Cu_3O_x$ single crystal with non-uniformly distributed twins. The mean induction $\bar{B}=14$ G. The photo is taken in the scanning electron microscope.

The above estimate is in good agreement with the pinning force determined using the induction gradients in the decoration patterns: $F_p = j_c \cdot B = \bar{B} \cdot \frac{dB}{dx} \simeq \bar{B} \cdot \frac{\Delta B}{\Delta x}$. We observed sharp local induction gradients in the regions with nonuniformly distributed twins as shown in Fig.4. For the samples with twin boundary spacings larger than $5 \cdot 10^{-5}$ cm we found the volume pinning force $F_p = 3 \cdot 10^6 \div 6 \cdot 10^7$ dyn/cm^3 and corresponding critical currents $j_c = (0.5 \div 1.5) \cdot 10^5$ A/cm^2 for the mean induction in the sample $\bar{B}=14$ G. In the sample with a lower twin density (as shown in Fig.3, twin boundary spacings $1.5 \div 2$ μm) we approximately estimated $F_p = (2 \div 5) \cdot 10^4$ dyn/cm^3 and $j_c = (2 \div 5) \cdot 10^3$ A/cm^3.

<u>Acknowledgements</u>. The authors are very grateful to L.I.

Burlachkov, M.V.Indenbom, L.A.Dorosinsky for useful discussions, to L.G.Isaeva for assistance.

References

1. J.G.Bednorz, K.A.Müller, 1986, Z.Phys., <u>B64</u>, 189.
2. U.Essmann, H.Träuble. Phys.Lett., 1967, <u>A24</u>, 596.
3. F.L.Gammel, D.J.Bishop, C.J.Dollan, J.R.Kwo, C.A.Murray, L.F.Schneemeyer, J.V.Waszczak. Phys.Rev.Lett., 1987, <u>59</u>, No.22, 2592.
4. L.Ya.Vinnikov, L.A.Gurevich, G.A.Emelchenko, Yu.A.Ossipyan. Soviet Pisma Zh.Eksp.Teor.Fiz., 1988, <u>47</u>, No.2, 109.
5. L.Ya.Vinnikov, L.A.Gurevich, G.A.Emelchenko, Yu.A.Ossipyan. Sol.St.Commun., 1988, <u>v.67</u>, No.4, 421.
6. G.A.Emelchenko, M.V.Kartsovnik, P.A.Kononovich, V.A.Larkin, V.V.Ryazanov, I.F.Schegolev. Sov.Pisma Zh.Eksp.Teor.Fiz., 1987, <u>46</u>, No.4, 162.
7. L.Ya.Vinnikov, A.O.Golubok. High-resolution technique of direct observation of magnetic structure in type-II superconductors. Preprint, Chernogolovka, 1984.
8. T.Forgan. Nature, 1987, <u>v.329</u>, 483.
9. A.I.Larkin, Yu.N.Ovchinnikov. JLTP, 1979, <u>v.34</u>, No.3/4, 409.
10. L.Ya.Vinnikov, I.V.Grigorieva. Sov.Pisma Zh.Eksp.Teor.Fiz., 1988, <u>47</u>, No.2, 89.
11. V.K.Vlasko-Vlasov, M.V.Indenbom, Yu.A.Ossipyan. Sov.Pisma Zh.Eksp.Teor.Fiz., 1988, <u>47</u>, No.6, 312.
12. L.I.Burlachkov. Vortex structure anisotropy in high-T_c superconductors. Submitted to Sov.Fiz.Nizk.Temp.
13. A.M.Grishin, A.Yu.Martynovich, S.V.Yampolsky. Preprint Don. FTI-88-12(149), 1988.
14. M.V.Kartsovnik, V.A.Larkin, V.V.Ryazanov, N.S.Sidorov. Sov. Pisma Zh.Eksp.Teor.Fiz., 1988, <u>47</u>, No.11, 595.
15. A.P.Volodin, M.S.Khaikin. Sov.Pisma Zh.Eksp.Teor.Fiz., 1981, <u>34</u>, 275.
16. A.Campbell and J.E.Evetts. Adv.Phys., 1972, <u>21</u>, 199.
17. M.Tinkham, Introduction to superconductivity, McGraw Hill, New York, 1975.

CRITICAL STATE AND FIELD H_{c1} IN $YBa_2Cu_3O_{7-x}$ SINGLE CRYSTALS

M.V.Kartsovnik (Institute of Chemical Physics, USSR Academy of Sciences), V.M.Krasnov, V.A.Larkin, V.V.Ryazanov, I.F.Schegolev (Institute of Solid State Physics, USSR Academy of Sciences, Chernogolovka, Moscow district, 142432, USSR)

The anisotropic properties of high-temperature superconductors can be studied only on single-crystal samples and oriented films. Typical YBaCuO single crystals obtained contain many defects and are of small transverse dimensions. This defines the speciality of measurements of anisotropic critical parameters /1,2/.

The purpose of this work is to study the anisotropy and also temperature dependences of the lower critical field H_{c1} and the critical density of current I_c in $YBa_2Cu_3O_{7-x}$ single crystals.

Magnetic measurements in fields of up to 400 G were made by means of rf SQUID magnetometer, the sensitivity of measuring the magnetic moment M being better than $10^{-8} A \cdot m^2$. Single crystals were grown from a non-stoichiometric melt of a mixture of

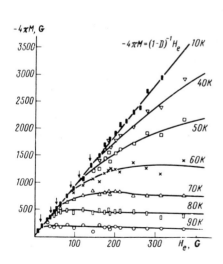

Fig.1. Dependence of the mean magnetization on the external magnetic field at different temperatures.

Arrows show the values of the external critical field $(H_e)_{c1}$.

Solid curves depict the calculation made by interpolation formula (1).

oxides /3/. The crystals were thermally treated in an oxygen atmosphere, which reduced essentially the width of their superconducting transition. Before and after annealing, a thick net of twins was detected in crystals when observed in a microscope in polarized light. Crystals are twinned in two mutually-perpendicular (110) planes. In most samples the regions twinned in different directions alternated and filled up uniformly the crystal.

To elucidate the influence of crystal sizes on dependences M(H) a $0.8 \times 0.9 \times 0.03$ mm^3 single crystal was cleaved after a series of measurements, and then the same investigations were carried out with one $0.5 \times 0.3 \times 0.03$ mm^3 fragment. Fig.1 shows a set of curves of the external magnetic field dependence of magnetization, $-4\pi M(H_e)$, for the fragment in the field $H_e \perp ab$. These dependences were obtained from the initial ZFC-curves of diamagnetic screening, $4\pi M(T)$, measured in a fixed field, turned on at T<10K. Fig.2 indicates ZFC-curves and Meissner pushing out (FC-curves) in the field H∥ab for a single crystal before and after annealing. A comparison with the lead reference sample suggests that the diamagnetic screening moment of single crystals at H∥ab with an accuracy of about 10% coincides with the ideal diamagnetic moment of lead. Therefore, for gauging $4\pi M$ in the case of $H_e \perp ab$ for the initial linear regions it was

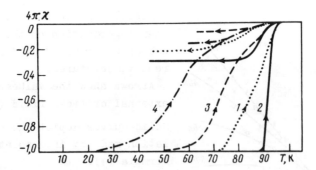

Fig.2. Temperature dependence of the magnetic susceptibility for diamagnetic screening (ZFC) and the Meissner pushing out in the field H∥ab: 1 - unannealed single crystal (H=5 G), (2-4) - the same crystal after annealing (H=5, 120 and 292 G).

assumed that $-4\pi M=H$ with $H=H_e/(1-D)$, where D is the demagnetization factor calculated on the assumption that the sample is of the ellipsoidal shape. For the single crystal fragment D= 0.893 ± 0.002, for the initial crystal D=0.946 ± 0.002.

Fig.3 presents temperature dependences of the lower critical field H_{c1} (characterizing the beginning of entry of the flow into the sample), which was determined by point $(H_e)_{c1}$ of deviation of dependence $4\pi M(H_e)$ from the initial linear behaviour (Fig.1). This point was chosen by using the interpolation curve $4\pi M(H_e)$ discussed below. Curves $H_{c1}^\perp(T)$ and $H_{c1}^{||}(T)$ for both the fragment and the initial single crystal are linear to the temperature of 40K, at a lower temperature it is hardly possible to get the value of H_{c1} from our result. A good agreement of the data obtained for different crystals suggests correct allowance made for the effect of their dimensions by means of a demagnetization factor. The ratio $H_{c1}^\perp/H_{c1}^{||} \simeq 6$. For the temperatures below 80K dependence $H_{c1}(T)$ differs essentially from that known for traditional superconductors: $H_{c1}\sim/1-(T/T_c)^4/$ (see also /4,5/). By extrapolating linearly $H_{c1}(T)$ to T=0 one may get lower estimates of the penetration depth $\lambda_{ab}(0)\gtrsim 5\times 10^{-6}$ cm and $\lambda_\perp(0)\gtrsim 3\times 10^{-5}$ cm (for $\varkappa_{ab}=25$).

Fig.1 indicates that curves M(H) for $60K \leq T \leq 90K$ run into the level $4\pi M_c(T)$, which is practically independent of H_e. As well as the authors of /1,2,6/, we attribute such a behaviour

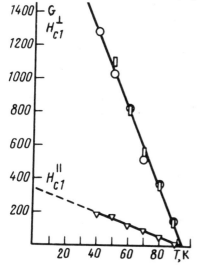

Fig.3. Temperature dependences $H_{c1}^\perp$ and $H_{c1}^{||}$. Different points for $H_{c1}^\perp$ correspond to the initial crystal and fragment.

to strong pinning of Abrikosov's vortices in the sample. The entering of a magnetic flow proceeds gradually with increasing the field at the sample edge /7,8/. As soon as the flow reaches the middle of the sample, the magnetic moment should stop increasing, the density of the superconducting screening current at any point of the sample being equal to the critical value of I_c (for small values of the field used in our experiments we assume I_c independent of H).

Fig.4 shows the temperature dependences of the magnetization critical current density for both the basal plane ($I_c^{\parallel}$) and the perpendicular direction ($I_c^{\perp}$). Values of $I_c^{\parallel}$ (the upper curve) were determined from the ratio /7/: $I_c = -30 M_c/R$, which is correct for a disc with radius R in perpendicular fields /2/. Here I_c is measured in A/cm^2, M_c in G and R in cm. The lower curve in Fig.4 presents the estimate of the critical current density in c-direction. To calculate numerically this value the base experimental data $M_c(T)$ for $H_e \parallel ab$ were used. In this case magnetic moment is created by both the currents in the basal planes and the currents across the Cu-O layers. It is possible to calculate $I_c^{\perp}$ if one assumes that the critical current magnitude $I_c^{\parallel}$ is the same both for $H \perp ab$ and for $H \parallel ab$. These calculations are not fully justified because the pinning force can be different in these cases.

We used following interpolation to determine more exact

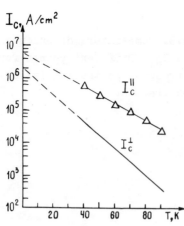

Fig.4. Temperature dependence of the magnetization critical current density in the basal plane (the upper curve) and the estimate of the dependence for c-direction.

values M_c and H_{c1} for T<60K. For a long cylinder in the field parallel to its axis mean magnetization of the sample is of the form /7/:

$$-4\pi M = H + (H^2 - H_{c1}^2)/12\pi M_c - (H_{c1}^2(3H - 2H_{c1}) - H^3)/432\pi^2 M_c^2 \quad (1)$$

Here H is the field intensity in the sample. In the case of a uniformly magnetized disc, that is, when the flow does not enter far into the sample, the substitution of $H = H_e/(1-D)$ into (1) is likely to be valid. The calculation of the exact formula for M under inhomogeneous magnetization of a thin disc in a perpendicular field is a separate problem which cannot be solved analytically in the general case. Besides, we can see from Fig.1 that with the above substitution (1) adequately interpolates the experimental data also when the magnetic flow enters rather far into the sample, in the range of 60K≤T≤90K at least. On the assumption that the interpolation formula (1) remains valid at temperatures below 60K, we can apply it for improving the low-temperature values of H_{c1} and determining the values of I_c in the range of 40-60K. As was the case in /9/, we have obtained the exponential dependence $I_c = I_{c0} \exp(-\alpha T)$, where $I_{c0}^{\parallel} \simeq 6 \times 10^6$ A/cm^2 and $\alpha^{\parallel} \simeq 6.2 \times 10^{-2}$ 1/K. (The correspondent values for the lower curve are $I_{c0}^{\perp} \simeq 2 \times 10^6$ A/cm^2, $\alpha^{\perp} \simeq 9 \times 10^{-2}$ 1/K) The temperature dependence $I_c(T)$ is also satisfactorily described by the power law $I_c \sim T^\nu$ with $\nu \simeq -3.5$.

Thus, from the magnetization curves we have obtained temperature dependences $H_{c1}(T)$ and $I_c(T)$ for the $YBa_2Cu_3O_{7-x}$ single crystals, which differ essentially from the corresponding dependences for traditional superconductors. The use of demagnetization factor has been shown to be effective way of taking account of finite dimensions of the crystals under estimation.

The authors are grateful to A.I.Larkin, Yu.N.Ovchinnikov for useful discussions, to G.A.Emel'chenko and N.S.Sidorov for affording the samples, to N.S.Stepakov for the assistance in adjusting the equipment.

1. Dinger T.R.,Worthington T.K. et al. //Phys. Rev.Lett. 1987. V.58.P.2687.

2. Avdeev L.Z., Bykov A.B. et al. //Pis'ma Zh. Eksp. Teor. Fiz. 1987. V.46. P.196.

3. Emel'chenko G. A., Kartsovnik M.V. et al. //Pis'ma Zh. Eksp.Teor.Fiz. 1987. V.46, P.16?.

4. Harshman D.R., Aeppli G. et al. //Phys.Rev. B. 1987. V.36. P.2386.

5. Rammer J. //Phys. Rev. B. 1987. V.36. P.5665.

6. Crabtree G.W., Liw J.Z., Umezawa A. et al. //Phys.Rev. B. 1987. V.36. P.4021.

7. Bean C.P. //Phys. Rev. Lett. 1962. V.8. P.250.

8. Bean C.P. //Rev.Mod.Phys. 1964. V.36. P.31.

9. Senoussi S., Oussena M., et al. //Phys.Rev.B. 1988. V.37. P.9792.

COPPER NMR AND NQR STUDIES OF THE $YBa_2Cu_3O_{7-x}$ HIGH T_c SUPERCONDUCTOR

E.Lippmaa, E.Joon, I.Heinmaa, A.Vainrub, V.Miidel and A.Miller (Institute of Chemical Physics and Biophysics, Tallinn 200001, Estonia, USSR),

I.F.Schegolev (Institute of Solid State Physics, Chernogolovka, USSR),

I.Furo and L.Mihaly (Central Research Institute for Physics, Budapest H-1525, Hungary)

Copper NMR and NQR spectra of the $YBa_2Cu_3O_{7-x}$ ($x = 0$ to 0.8) perovskite compounds have been studied and unequivocally interpreted. The principal components of the Knight shift and electric field gradient tensors have been determined and the spin-lattice relaxation has been studied for orthorhombic crystals in the normal state. The spin-lattice relaxation of Cu(2) nuclei in planes has been shown to be magnetic in origin and temperature independent for $c \parallel H$ orientation. In case of $c \perp H$ the relaxation is magnetic predominantly, the relaxation rate slightly increasing with the temperature, with the exception of a narrow region near 220 K where the quadrupole contribution seems to be important. On the other hand, the quadrupole relaxation is predominant for Cu(1) nuclei in chains.

Line Assignments and Interaction Parameters in the Orthorhombic Crystals

There are two different copper sites in the $YBa_2Cu_3O_{7-x}$ structure: the fivefold oxygen-coordinated Cu(2) in planes and Cu(1) in chains which are predominantly fourfold oxygen-coordinated in the orthorhombic crystals (x = 0 to 0.6) and predominantly twofold oxygen-coordinated (x = 0.6 to 1) in the tetragonal oxygen-depleted ones. The two copper NQR lines in the spectra of orthorhombic samples at approximately 22.2 MHz and 31.2 MHz were at first assigned according to the relative line intensities to the Cu(2) and Cu(1) sites, respectively /1/. Although this assignment was supported by preliminary NMR evidence /2/, it had to be reversed after careful study of copper NMR lineshapes /3, 4, 5/, influence of Y substitution by Gd /6/ and, most convincingly, from single crystal studies /5, 6, 7/.

The NQR and NMR study of a $x \cong 0$ powder sample /3/ provided fairly accurate Knight shift and quadrupole coupling data, which were used for the interpretation of the angle dependence of single crystal NMR spectra /7/. The angle dependence was measured in 8.5 and 11.7 Tesla fields using one twinned 2.2 x 2.0 x 0.4 mm^3 (8 mg) x $\approx$ 0, T_c = 91 K and ΔT_c = 1.8 K single crystal and a mosaic (28 mg) of similar twinned x $\approx$ 0 crystals. The ^{63}Cu NMR spectrum at 100 K of the mosaic, oriented with the c-axis parallel to the magnetic field direction, is shown in Fig. 1a. In accordance with the known ^{63}Cu NQR frequencies /3/, ν_{Q1} = 22.2 MHz for Cu(1) in the chains and ν_{Q2} = 31.2 MHz for Cu(2) in the planes, and the electric field gradient tensor asymmetry parameters $\eta_{Q1} \approx 1$ and $\eta_{Q2} \approx 0$, estimated from the lineshapes of ^{63}Cu NMR spectra of polycrystalline samples, we can assign these three lines as follows: a is the central (1/2 $\leftrightarrow$ -1/2) transition of Cu(2), b is the central transition of Cu(1) and c is the satellite ($\pm$3/2 $\leftrightarrow$ $\pm$1/2) transition of Cu(1). This assignment is confirmed by the angle dependence of the frequencies of the two central transition lines, presented in Fig. 2. The intensity ratio of these lines is 1:2, as is seen from Fig. 1b, where the satellite transition has already vanished due to excessive broadening. This spectrum was registered using only one 8 mg single crystal. The weaker line is also broader because of directional disorder in the a-b plane due to twinning (250 kHz at half height instead of 80 kHz). These lines do not show any splittings and thus $\eta = 1$ and $\eta = 0$ is a better approximation than $\eta = 0.95$ and $\eta = 0.15$ as was deduced from the powder spectra /3/.

The central transition angle dependences were fitted by the formula /8/

$$\nu = \nu_0 \left[(1 + K_{xx})\sin^2\theta \sin^2\varphi + (1 + K_{YY})\sin^2\theta \cos^2\varphi + (1 + K_{zz})\cos^2\theta \right] - (V_{zz}^2/2\nu_0) \left[A(\varphi)\cos^4\theta + B(\varphi)\cos^2\theta + C(\varphi) \right],$$

where ν_0 is the Larmor frequency, θ, φ are the angles of the magnetic field H_0 as referred to the principal axis system of the field gradient tensor, K_{ij} - the Knight shift and $V_{ij} = (eQ/2h)\partial V/\partial x_i \partial x_j$ the electric field gradient tensors,

$$A(\varphi) = -27/8 - (9/4)\eta\cos 2\varphi - (3/8)\eta^2\cos^2 2\varphi,$$

$$B(\varphi) = -30/8 - (\eta^2/2) + 2\eta\cos^2\varphi + (3/4)\eta^2\cos^2 2\varphi,$$

$$C(\varphi) = -3/8 + (\eta^2/3) + (\eta/4)\cos 2\varphi - (3/8)\eta^2\cos^2 2\varphi,$$

describing the resonance condition of the $I = 3/2$ nuclei in the presence of the second order quadrupolar and anisotropic Knight shift interactions. This leads to the following parameters for the electric field gradients and the Knight tensors /7/: $K_{aa}=1,34\%$; $K_{bb} = K_{cc} = 0.57\%$; $V_{aa} = \pm 19.2$ MHz, $V_{bb} = \mp 19.2$ MHz, $V_{cc} = 0.0$ MHz for Cu(1) and $K_{aa} = K_{bb} = 0.57\%$, $K_{cc} = 1.27\%$; $V_{aa} = V_{bb} = -15.6$ MHz, $V_{cc} = 31.2$ MHz for Cu(2). It is evident that the c-axis is a good axial symmetry axis for both Cu(2) tensors, but the axial symmetry of the Cu(1) Knight shift in the chains as well as the extreme asymmetry of this electric field tensor are rather unexpected. These results are in a reasonably good agreement with the data of Slichter /9/, but in disagreement with the results of Horvatic et al /10/.

Line Assignment and Oxygen Ordering in the Tetragonal Compound

Only one line is present in both the NQR and NMR spectra of copper in the antiferromagnetic tetragonal ($x > 0.5$) compound. The ^{63}Cu NQR resonance appears at 30.1 MHz for $x = 0.8$ and it is tempting to assign it to Cu(2) because of the closeness of resonance frequencies and lineshapes, but it would be in contradiction with the neutron scattering results of Shirane /12/. No unambiguous proof is possible with the yttrium compound and the question was left open for further study /11/. We have now prepared a polycrystalline sample of $TmBa_2Cu_3O_{6.4}$. Since the Tm^{3+} magnetic moments create dipole fields of different signs at the Cu(1) and Cu(2) sites, comparison of the high field ^{63}Cu NMR spectra of polycrystalline tetragonal Y and Tm compounds provides unambiguous assignment of the copper sites just as was the case with the superconducting Gd compound studied by

Kitaoka /6/. The calculated dipole fields, involving contributions from the lanthanide magnetic moments within a R = 20 Å sphere for Cu(2) and a R = 25 Å sphere for Cu(1) are in very good agreement with the calculations of Kitaoka.

The observed 0.7 MHz shift of the x = 0.6 thulium compound ^{63}Cu NMR powder pattern high frequency edge (corresponding to crystalline orientation c H) to a lower frequency (see Fig. 3) immediately assigns this resonance to Cu(1) sites in the chains, in accordance with the neutron scattering data. The presence of localized magnetic moments at the Cu^{II} sites in the plane makes it unobservable by NMR and NQR. It also follows that no such Cu^{II} localized moments are present in the chains even though the formal valence state of copper passes here from Cu^{III} to Cu^{I} with oxygen depletion.

Relaxation

The ^{63}Cu NQR and NMR relaxation processes have been studied over a broad temperature range by many authors. The total ^{63}Cu relaxation rate $1/T_1$ generally increases with increasing temperature with a slight discontinuity around 220 K /1/. The Cu(2) NQR relaxation rate temperature dependence in the normal state of the orthorhombic phase is rather weaker than that of Cu(1) in the chains /13/ and the mechanism is considered to be predominantly magnetic. In most studied the experimental data were fitted to a single exponent, or if this failed, only the slow tail of the magnetization recovery was taken into account. Actually relaxation of a quadrupolar nucleus is a multi-exponential process and for the central transition of the spin 3/2 (copper) nucleus is described by the sum of two exponents

$$M(\infty) - M(\tau) = Ae^{-\tau/T_{11}} + Be^{-\tau/T_{12}},$$

with $T_{12}/T_{11} = 6$ for a purely magnetic interaction /14/. Taking into account the axial symmetry of the electric field gradient tensor it should be the case for $c \parallel H$ orientation. T_{12}/T_{11} ratio is different in case of other, non-magnetic interactions.

The spin-lattice NMR relaxation process measured in 8.5 T magnetic field for the central transition of the Cu(2) sites in the oriented mosaic of $x \approx 0$ orthorhombic single crystals is indeed well described by the sum of two exponents with $T_{12}/T_{11} = 6$ in the whole temperature range for case of $c \parallel H$ (Fig. 4). Therefore, it is purely magnetic in origin, as it should be. Surprisingly, the relaxation rate is independent on the temperature (Fig. 5). The $T_{12}/T_{11} = 6$ ratio is also characteristic of $c \perp H$ orientation for the most part of the temperature interval with the exception of a narrow region near 220 K, where the slow component of the relaxation rate exhibits a sharp maximum (Fig. 5). This anomaly seems to be associated with the quadrupole contribution into the relaxation rate that may reflect some peculiarities in lattice dynamics. Raman spectroscopy reveals temporary formation of CuO_6 octahedra in this temperature range /15/.

The spin-lattice relaxation of the Cu(1) nuclei may be also described by the sum of two exponents, however the ratio T_{12}/T_{11} is far from 6 here. Therefore, the role of the quadrupole contributions is essential.

The absence of any temperature dependence in the relaxation rate of Cu(2) sites reveals convincingly the absence of the conductivity electrons on the copper atoms in the plane. The temperature independent magnetic relaxation may be an evidence of the fact that the Cu(2) electrons form a singlet state magnetic system with the characteristic exchange energy much higher than, say, 300 K.

References

1. I.Furo, A.Janossy, L.Mihály, P.Banki, I.Pocsik, I.Bakonyi, I.Heinmaa, E.Joon, E.Lippmaa, Phys.Rev.B, No.36, 5690 (1987).
2. W.W.Warren Jr., R.E.Walstedt, G.F.Brennert, G.P.Espinosa, J.P.Remeika, Phys.Rev.Lett., v.59, 1860 (1987).
3. E.Lippmaa, E.Joon, I.Heinmaa, V.Miidel, A.Miller, R.Stern, I.Furo, L.Mihály, P.Banki, Physica C, 153-155, 91 (1988).
4. M.Mali, D.Brinkmann, L.Pauli, J.Roos, H.Zimmermann, Phys. Lett. A124, 112 (1987).
5. H.Lütgemeier, M.W.Pieper, Sol.St.Comm., 64, 267 (1987).
6. J.Kitaoka, S.Hiramatsu, K.Ishida, T.Kohara, Y.Oda, K.Amaya, K.Asayama, Physica B, 148, 298 (1987).
7. I.Heinmaa, A.Vainrub, J.Past, V.Miidel, A.Miller, I.Schegolev, G.Emelchenko, V.Tatarchenko, Pisma JETF, 48, 171 (1988).
8. J.F.Baugher, P.C.Taylor, T.Oja, P.J.Bray, J.Chem.Phys., v.50, 4914 (1968).
9. C.H.Pennington, D.J.Durand, D.B.Zax, C.P.Slichter, J.P.Rice, D.M.Ginsberg, Phys.Rev. B37, 7944 (1988).
10. M.Horvatic, D.Ségransan, Y.Berthier, P.Butaud, J.Y.Henry, M.Couach and J.P.Chaminade, Phys.Rev.Lett., submitted.
11. L.Mihaly, I.Furo, S.Pekker, P.Banki, E.Lippmaa, V.Miidel, E.Joon, I.Heinmaa, Physica C, 153-155, 87 (1988).
12. J.M.Tranquada, D.E.Cox, W.Kunnmann, H.Moudden, G.Shirane, M.Suenga, P.Zollikev, D.Vaknin, S.K.Sinha, M.S.Alvarez, A.J.Jacobson, D.C.Johnston, Phys.Rev.Lett., 60, 156 (1988).

13. H.Yasuoka, T.Shimizu, T.Imai, S.Sasaki, Y.Ueda, K.Kosuge, Tech.Rep. of ISSP A No. 1998, July (1988).
14. A.Narath, Phys.Rev., 162, 320 (1967).
15. L.A.Rebane, T.A.Eimberg, E.M.Fefer, G.E.Blumberg, E.R.Joon, Sol.St.Comm., 65, 1535 (1988).

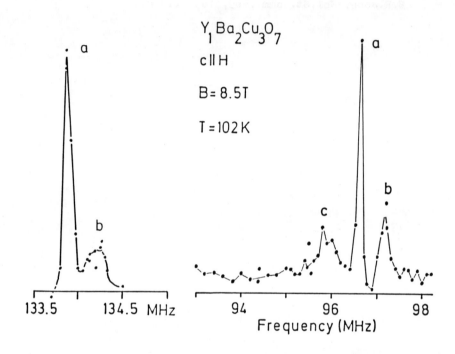

Fig. 1. ^{63}Cu NMR spectrum of c ∥ H oriented aligned single crystals (28 mg) at 8.5 T field at 102 K (a) and of a single crystal (8 mg) at 11.7 T field oriented at a = CH = 10.2° at room temperature.

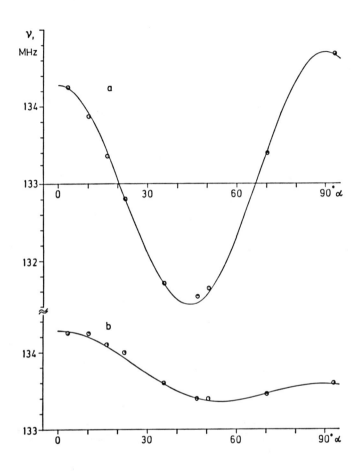

Fig. 2. The angle dependences of the frequencies of the lines a and b. The spectra of the single crystal were recorded at 11.7 T field at room temperature.

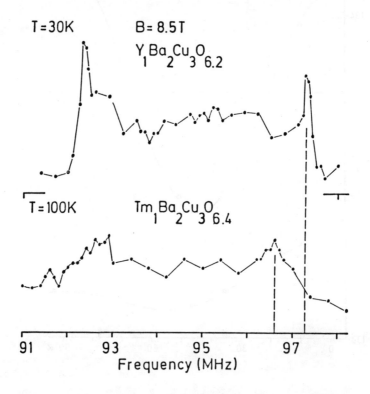

Fig. 3. The ^{63}Cu NMR spectra of tetragonal YBa$_2$Cu$_3$O$_{6.2}$ and TmBa$_2$Cu$_3$O$_{6.4}$.

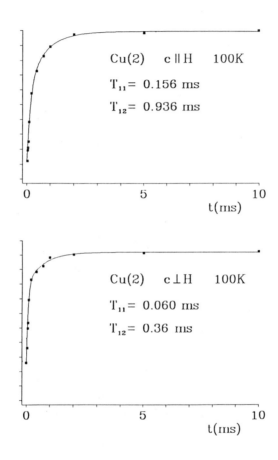

Fig. 4. Examples of the magnetization recovery. Solid line corresponds to the two-exponential fit.

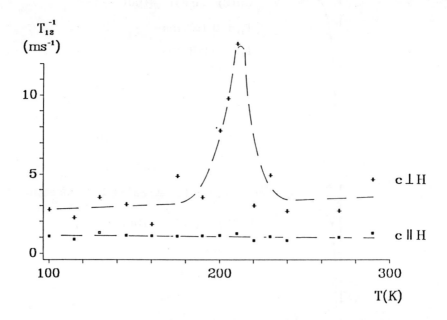

Fig. 5. The temperature dependence of the relaxation rate at two orientations: dashed line - a guide to the eye.

SCANNING TUNNELING SPECTROSCOPY OF A HIGH-T_c SUPERCONDUCTING MONOCRYSTAL $YBa_2Cu_3O_7$

A.P.Volodin[I], M.S.Khaikin, G.A.Stepanyan[I]

[I] Institute for Physical Problems Acad. Scie., USSR, Moscow
Moscow Institute of Steel and Alloys

ABSTRACT

The energy gap distibution on the surface of the yttrium ceramic monocrystals were measured by STM. Microcrystals of some tens μm dimentions were grown in ceramic during sintering cycle. The gap values are in range $(2.4 \div 7)\ 2\Delta/kT_c$. The measurements were done at 4K within the area of 1 μm^2 in 110 points spaced by 100 nm. In each point the tip penetrated into the nonconducting surface layer breaking it till obtaining the tunneling current of 1 nA. After that the tip moved backward and toward the crystal again till obtaining the vacuum tunnel current of 1 nA. Just so were achieved the reproducible conductance - voltage characteristics for the gap measurements. In such a way the space diagrams of the gap distribution were performed. One of the diagrams showed both superconducting and normal regions of a sample. The nature of nonconductive surface layer is discussed. As can be seen by comparing of topogramms the exposition in vacuum leads to variations in surface conductivity.

Tunneling spectroscopy is one of the most powerful tools for the study of the superconductivity. It can provide a direct information about the energy gap and phonon spectra.

The scanning tunneling microscope (STM), which affairs a resolution at the atomic level has made it possible to carry out several remarkable studies of the topography of conducting samples. Scanning tunneling spectroscopy (STS), i.e., STM as a spectroscopic tool, has many merits compared with traditional tunneling spectroscopy. The vacuum provides the ideal barrier for tunneling that was introdused in STM. STM provides spectroscopic images with atomic spatial resolution. STS is obviously of major importance for research on superconductivity of high-T_c superconductors[1]. STS measurements, for example, may give insight in local superconductivity at grain or twin boundaries[2].

de Lozanne et al.[3] obtained profiles of electronic properties of a Nb_3Sn film, including a transition from a superconducting region to a normal region at 4 K. They also obtained current-voltage characteristics of a tunnel junction between the needle of a scanning tunneling microscope and the superconducting sample in selected points.

It did not prove possible to use STM directly to study high-T_c superconductors: a surface layer of a high-T_c superconductor at low temeratures goes nonconducting. As a result, when the needle of STM is brought in contact with the sample to obtain a tunnel current, it touches the nonconducting layer.

In experiments[4,5] STM was used to study high-T_c superconductors at low temperatures. Current-voltage characteristics were obtained in various points of samples. These characteristics revealed significant differences in the values of energy gap from point to point. The tunnel current apparently flowed across a thin nonconducting layer and was accompanied by a parralel leakage current. Scanning was not carried out for the reason stated above and also because the particular STM which were used had a scanning amplitude of only ~1000 Å at liquid-helium temperature, or roughly equal to the thickness of the needle tip.

In this article we will describe measurements of energy gap and its spatial distribution in the high-T_c microcrystals. The gap measurements were made with a low-temperature bimorf-elements STM described briefly in ref.6. This STM operates under vacuum at 300 K and in liquid helium at 4 K. The tip used was made of W monocrystal etched in NaOH solution.

The sample was a ceramic $YBa_2Cu_3O_7$ containing single crystals with sizes 20-60 μm, with a natural shiny surface and having the value $T_c=92\pm2$ K. Samples was prepared by traditional technology of high-temperature sintering process. Before being investigated, part of the sample was cleaved in order to produce a surface as fresh as possible. Crystal faces suitable for study were sought by recording "coarse" topograms of the sample with a woltage $V\sim5$ volts on the needle. The occurence of a varying degree of oxigen depletion in surface layer was confirmed by the fact that vacuum conditions changed the conductivity of a surface layer.

The following experiment was undertaken in order to find out the influence of the vacuum conditions on the sample's surface: topograms of one and the same part of the surface of the sample that had never been under vacuum conditions were recorded on the first minute after it was placed in vacuum with pressure of the oxigen $\sim 10^{-6}$Torr (fig.1a) and 12 hours later (fig.1b). Recording was carried out with needle-sample voltage V=5 volts, which gave the best results of scanning. Reproducebility of the scans, lack of noise and outliers is usual for the regime of vacuum tunneling. Optimum value of the voltage indicates the semiconducting type of the surface layer.

As can be seen by comparing of topograms (fig.1a,b), exposition in vacuum leads to remarkable changing in recorded topogramms and increasing of noise. The reproduce bility of the scans decreased for 12-hours vacuum-exposed sample. Difference between fig. 1a and fig. 1b can be explained by forming of thin insulating layer on the sample surface in vacuum. In this case STM works in the regime of mechanical contact between the needle and the surface which requires forsing of needle against the surface to provide observable current. This leads to piezoelement-controlling amplifier reading-off scale, that results in steps on topogram (fig.1b). Dielectric layer on sample surface is formed, perhaps, by the escape of mobile atoms of oxigen from surface layer which is in dynamical equilibrium with an atmospheric oxygen[7].

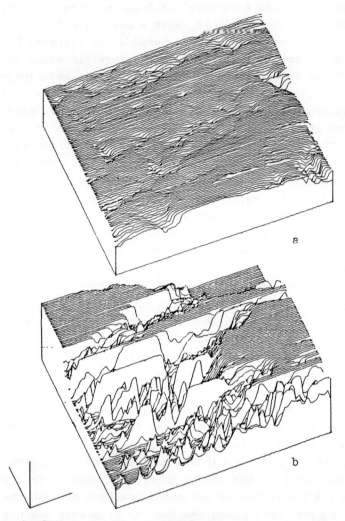

FIG. 1a. An STM topogram of $YBa_2Cu_3O_7$ surface at room temperatures in first minutes of exposition in vacuum chamber. Divisions on the axis corresponds to 2500 Å.
FIG. 1b. An STM topogram of the same surface at room temperature after 12-hour exposition in vacuum chamber. Divisions on the axis corresponds to 2500 Å.

The tunnel current from the needle to the point of the sample under study, I_t, at low temperature was produced by the following procedure[8]. At V ~100 mV the needle was brought close to the sample (at a point which was not touched) and immersed to a depth $z \sim 200$ Å (determined from the piezoelectric drive) into the nonconducting layer, until a current $I_t \sim$ 1nA was reached. We concluded that the needle made contact with the solid surface when the characteristic vibrational noises in I_t disappeared. The current-voltage charactiristics obtained with the needle in this position were highly asymmetric; furthermore, they contain a significant parabolic component (a leakage current; fig.2a). (Similar structural features were observed on I-V characteristics in Refs.4,5.) The $I_t(z)$ profile recorded upon the first contact of the needle with the sample at V=200 mV has the shape shown in fig.2b. The shape of this curve is quite different from the steep exponential curve characteristic for a tunnel current.

The needle was then withdrawn from the sample to $z \sim 200$ Å and then brought back close to the sample, until a current $I_t=$ =1 nA was reached. A needle displacement $z \sim$ 100 Å (according to the piezoelectric drive) was sufficient for this purpose. The apparent reason for the difference is a deformation of the needle and the sample during the first contact. The I-V characteristic and I(z) now acquired the form characteristic of a tunnel current across a vacuum gap (fig.2, c and d). The noise in I_t was of usial nature.

A repetition of this operation (at the same point of the sample) caused no further significant change in such characteristic of the contact as character of dependence dI/dV(V) and its first maxima (from V=0) for the derivative. Variations in averidge tunnel current I_t by the power (correspondingly vacuum barrier changed by 2-3 times) also did not lead to noticeble changings of maxima positions. It indicates an absence in the first approximation of effects like " Coulomb blokade" of tunneling, caused by charging of a small electric capacitance tip-surface by tunneling of electrons[9].

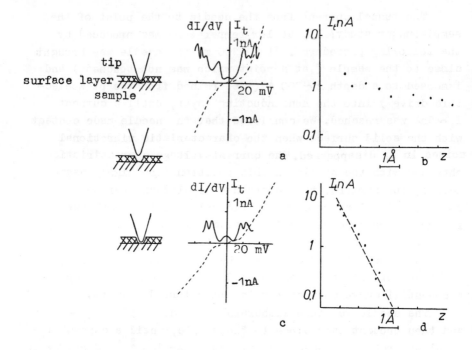

FIG. 2. a - Current-voltage characteristic of the first contact of the needle with the sample (dashed line) and its derivative (solid line); c - current-voltage characteristic and its derivative during repeated contact; b - current versus distance from the needle to the sample during the first contact d - the same, during repeated contact. A left part of fig. indicates the tip position during the measuring procedure.

The apparent explanation of these observations is that when the needle makes first contact, an oxygen-poor, nonconducting surface layer of the sample crumbles and is removed, exposing the superconducting part of the sample. There is no further loss of oxygen at liquid-helium temperature. When it is subsequently moved toward the sample, the needle approaches the surface of the superconductor without hindrance.

The apparent reason for the observed at first contact dependence I_t (fig. 2b) caused by existing of nonconductive hindrance between the needle and surface. Formed by this procedure vacuum barrier exclude a pressure effect and, hence, possible distortion in voltage-current characteristics.

Distributions of the energy gap along the surface of a sample were measured by following technique. The current-voltage characteristic of the tunnel junction at the given point on the sample was differentiated (fig.2c), and the distance between the maxima of the derivative was stored as a measure of the energy gap. The needle was then raised to $z \sim 200\text{Å}$ and moved above the surface of the sample to the next point. The procedure for producing a tunnel contact with a characteristic of the type fig. 2c was then repeated. This procedure was used to over the entire surface area of the crystal under study, at 11x10 points, at steps of 1000 Å. The steps could not be made smaller because of the resulting instabilities and scatter in the measurements, apparently because the needle tip, 1000 Å in diameter, struck a part of the crystal surface which had been damaged in the preceding step.

Fig.3 shows dI/dV data at two different position on the surface. The curves dI/dV(V) shows peaks at diffe-

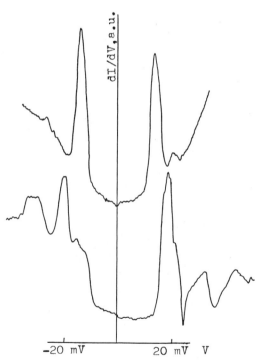

FIG. 3. dI/dV data for $YBa_2Cu_3O_7$ microcrystal at different points with a distance 1000 Å.

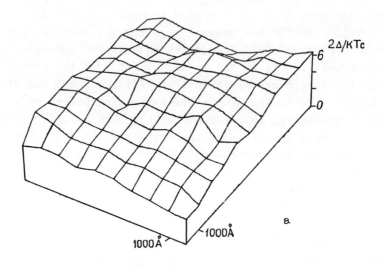

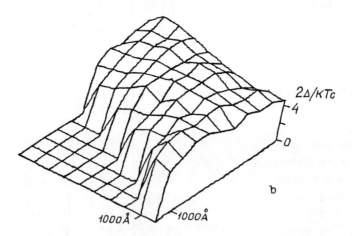

FIG. 4. Distribution of the energy gap. a – Over the surface of a supercunducting microcrystal; b – Over a part of a sample containing superconducting and normal regions.

rent energies at different points of the sample surface. The peaks in spektra outside the gap region change their position more drastically. An interpretation of the origin of these peaks seems difficult.

Fig. 4a and 4b shows results of the measurements of the initial gap in two areas with sizes of $1 \mu m^2$. The size of the gap, expressed in units of $n=2\Delta/kT_c$, is shown by the height of the points above the mesh point of the x,y coordinate system Points are connected by straight-line segments to help depict the n(x,y) surface. In fig. 4a, the values of n range from 2.4 to 6.3; those in fig. 4b range from 2.5 to 7; the error in the measurements is $n \sim 1$. The boundury between a superconducting crystal and a normal region of the sample falls in the field of fig. 4b; the tunneling current-voltage characteristic here is a straight line, so we have n=0.

In fig. 4,a and b, we can clearly see valleys and ridges running along the n(x,y) surfaces. This reproducibility of the results of measurements on several scan lines of the needle along the sample indicates that this measurements procedure is reliable. It also indicates a definite relationship between the size of the gap and structural features of the superconducting crystals; these features should be studied in parallel by other methods involving scanning tunneling microscopy and scanning tunneling spectroscopy.

However, such studies will require developing a method for exposing the surface of the superconductor or saturating the depleted surface layer with oxygen, in order to make it possible to use the highest resolution of the scanning tunneling microscope.

The samples studied in these experiments were prepared by N.M.Kotov, A.O.Komarov (of the Moscow Institute of Steel and Alloys), whom the authors deeply thank.

A.A.Abrikosov and A.S.Borovik-Romanov are thanked for their interest in this study.

1. J.G.Bednorz and K.A.Müller, Z.Phys. B64,189 (1986).
2. M.S.Khaikin and I.N.Khlustikov, JETP Lett.,33,167,(1981).
3. A.L.de Losanne, S.A.Elrod, C.F.Quate, Phys. Rev. Lett. 54,2433 (1985).
4. J.R.Kirtley, C.C.Tsuei et al., Jpn.J. Appl. Phys., 26, 997 (1987).
5. K.W.Ng, S.Pan, A.L. de Losanne et al., Jpn. J. Appl. Phys. 26,993 (1987).
6. M.S.Khaikin, A.P.Volodin, A.M.Troyanovskii, and V.S.Edel'-man. Prib. Tekh. Eksp. №4, 231 (1987).
7. L.E.C. van de Leemput, P.J.M. Van Bentum, et al.,Physica C152,99 (1988).
8. A.P.Volodin and M.S.Khaikin, JETP Lett., 46,588 (1987).
9. P.J.M. van Bentum, H.van Kempen et al., Phys. Rev. Lett., 60,369,(1988).

REFLECTIVITY AND RAMAN SPECTRA OF SUPERCONDUCTING
AND NONSUPERCONDUCTING $YBa_2Cu_3O_{7-\delta}$ CRYSTALS

M.P.PETROV, A.I.GRACHEV, M.V.KRASINKOVA, N.F.KAR-
TENKO, A.A.NECHITAILOV, V.V.POBORCHII, V.V.PROKO-
FIEV, S.S.RUVIMOV, S.I.SHAGIN

A.F.Ioffe Physical-Technical Institute of the Academy of Sciences, Leningrad, 194021, USSR

ABSTRACT

Preparation and optical studies of essentially single-domain and twinned (in the a-b plane) $YBa_2Cu_3O_{7-\delta}$ crystals are reported. Two types of anisotropy of reflection in the a-b plane have been observed; the first is attributable to the crystal symmetry, the other is governed by symmetry of twins. The experimental evidence for the effect of microtwinning on Raman spectra has been obtained.

INTRODUCTION

The experimental studies of physical properties of $YBa_2Cu_3O_{7-\delta}$ entered a qualitatively new stage when single crystals of this compound were produced. First of all, the possibility to obtain information on anisotropy of electrical, optical and other properties of $YBa_2Cu_3O_{7-\delta}$ appeared /1-4/. The properties of single crystals parallel and perpendicular to the c axis are easily determined, while the anisotropy in the a-b plane is usually masked by microtwinning /2,5/.

The paper discusses the experimental studies of orthorhombic $YBa_2Cu_3O_{7-\delta}$ crystals with ~ 100-μm single-domain regions in which we determined for the first time, to our knowledge, the anisotropy of reflection in the a-b plane. We have also found how twin boundaries affect the reflectivity and Raman spectra of twinned samples. Additional information was obtained in studies of reflectivity spectra of nonsuperconducting tetragonal $YBa_2Cu_3O_{7-\delta}$ crystals.

PREPARATION OF SAMPLES AND THEIR STRUCTURE

Superconducting $YBa_2Cu_3O_{7-\delta}$ crystals were obtained from the powder having the same composition and exhibiting superconductivity by spontaneous crystallization at 1000°C in oxygen. The crystals were thin plates of regular geometrical shape with mirror surfaces 200 x 300 μm^2 in size and 5-8 μm thick.

The powder diffraction patterns of single crystals are in good agreement with the X-ray data for the orthorhombic phase of $YBa_2Cu_3O_{7-\delta}$. The crystalline structure is well formed, the c axis is perpendicular to the plate plane, and the a and b axes coincide with natural faces of the crystal. The elementary cell parameters averaged for several single crystals are a = 3.821(8), b = 3.894(7), and c = 11.709(8) Å. According to our estimates, the content of oxygen in the crystals was $\delta \gtrsim 0.1$.

The superconducting properties of the samples were determined by the Meissner effect at temperatures up to 90 K.

As noted above, the $YBa_2Cu_3O_{7-\delta}$ crystals were twinned in the a-b plane, with the (110) twinning planes. Single domains ranged typically from 100 to 1000 Å in size, however, domains reaching $\sim$ 100 μm in size were found in our crystals, similar to $EuBa_2Cu_3O_{7-\delta}$ /3,7/. The domain structure of these crystals is well seen in a polarizing microscope (Fig.1). The bright regions correspond to separate domains which become bright in this geometry due to polarization plane rotation of incident light resulting from different reflectivities (R) for waves with E parallel to the a and b axes. Because the microtwins are optically unresolvable, the regions with microtwins behave here as optically isotropic and remain dark.

The transmission electron microscopy (TEM) studies of both types of regions confirmed the optical observations. A JEM-7A device with an accelerating voltage of 100 kV was used. The TEM-samples were prepared by ion thinning of $YBa_2Cu_3O_{7-\delta}$ crystals fixed on a mesh. This allowed the orientation of the

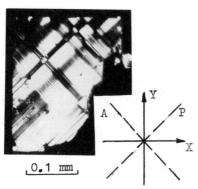

Fig.1 Polarized optical microscope photograph of $YBa_2Cu_3O_{7-\delta}$ single crystal in normal reflection mode. The polarizer (P) and analyzer (A) are orthogonal and the a and b axes of the sample (denoted by X and Y) are set at 45° to the polarizer direction.

a and b axes of the crystalline lattice in separate single-domain regions to be defined from an electron diffraction pattern.

Both the (110) and ($1\bar{1}0$) twins were observed, the sizes of separate domains of the orthorhombic phase ranged from 500 Å to ~ 100 μm. Figs.2 and 3 show electron microphotographs of these domains and their corresponding electron diffraction patterns. Both types of twins are seen in Fig.2. In Fig.3 two single domains of identical orientations separated by a narrow twinned region are seen.

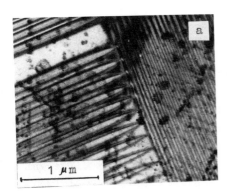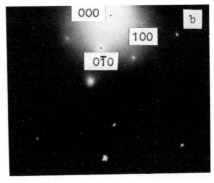

Fig.2 Electron micrograph of a microtwinned region of the crystal (a) and the corresponding electron diffraction pattern (b).

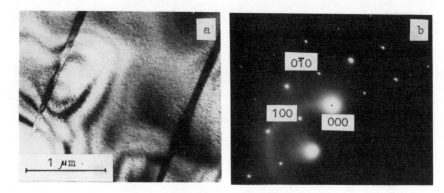

Fig.3 Electron micrograph of a nearly single-domain region of the crystal (a) and the corresponding electron diffraction pattern (b).

REFLECTIVITY SPECTRA

Polarized reflectivity spectra were recorded for orthorhombic crystals within the 1-3-eV range. The microphotometric technique which allows reflectivity from 10-50-μm regions to be measured was employed.

It was defined how the crystallographic a and b axes are directed relative to the crystal habit for several nearly single-domain samples. Therefore in further studies we could relate the behaviour of the reflectivity spectra to the direction of the polarization plane of incident light with respect to the crystal axes.

Fig.4a presents typical reflectivity spectra R_a (E ∥ a) and R_b (E ∥ b) for a single-domain region of the sample. The low-frequency rise of reflectivity exhibited by both R_a and R_b is consistent with the data obtained previously for ceramic and single-domain samples /5,8-10/. However, as distinguished from the data reported in /8/, R_a and R_b behave in different fashions and the reflectivity rise begins at different frequencies.

Measurements of reflectivity with decreasing temperature below room temperature have revealed that the plasma edge position remains the same for both polarizations, but the edge be-

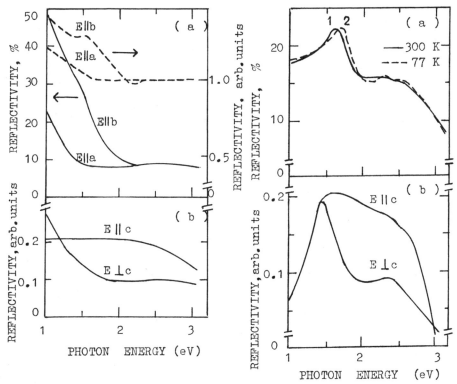

Fig.4 Reflectivity spectra of orthorhombic $YBa_2Cu_3O_{7-\delta}$ crystal. (a) Spectra in the a-b plane of a single-domain region at 300 K and spectra at 77 K normalized to room-temperature ones. (b) Reflectivity spectra for the c-axis-containing plane at 300 K. (E is the electric field vector of incident light wave).

Fig.5 Reflectivity spectra of tetragonal $YBa_2Cu_3O_{7-\delta}$ crystal. (a) Reflection in the (001) plane at 300 K and 77 K. (b) Reflection in the c-axis-containing plane at room temperature

comes sharper (Fig.4a, curves 3,4). At the superconductivity transition ($T_c \sim 90$ K), no jumps in the curves were observed.

Fig.4b shows reflectivity spectra obtained from the crystal edge. Because of a small thickness of the crystal, the absolute value of R is difficult to determine, however the behaviour of the curve is estimated reliably enough (which is demonstrated by the curve for $E \perp c$). For $E \parallel c$, reflectivity in the range studied changes only slightly and there is no indication of the plasma edge.

In addition to orthorhombic crystals, reflectivity spectra from nonsuperconducting crystals of the tetragonal phase with the parameters of an elementary cell $a = 3.875(2)$, $c = 11.740(6)$ Å were studied.

The reflectivity spectrum for the plane normal to the c axis is presented in Fig.5. It is characterized by well-pronounced reflectivity peaks at 1.6 and 2.25 eV, which shift towards shorter wavelengths with decreasing temperature (Fig.5, curve 2). Comparison of the spectra shown in Figs.4a and 5a reveals that the reflectivity peaks for the tetragonal phase and specific features of the reflectivity spectra of the orthorhombic phase observed at ~ 2.5 eV and ~ 1.5 eV are probably related to each other.

As far as the reflectivity spectra of the tetragonal crystals in the c-axis-containing plane are concerned (Fig.5b), a shift of the reflectivity peak at 1.6 eV towards longer wavelengths should be noted. This is attributable to spatial dispersion which is usually fairly strong in such anisotropic crystals as $YBa_2Cu_3O_{7-\delta}$.

The above results for orthorhombic crystals were obtained for single-domain regions. The study of reflectivity from regions with twins revealed the presence of another type of reflectivity anisotropy which arises because of symmetry of twins. The direction of one of principal axes in this case coincides with the boundary of twins. Since there are two equivalent directions of twinning planes, i.e., (110) and ($1\bar{1}0$), there are also two types of regions which are readily distinguished in polarized light. (Fig.6). The origin for reflectivity anisotropy in this case is illustrated by the spectral dependences shown in Fig.7. While the behaviour of the curve for $E \parallel y'$ parallel to the boundaries (i.e., the [110] or [$1\bar{1}0$] axis) is similar to that of the curve obtained

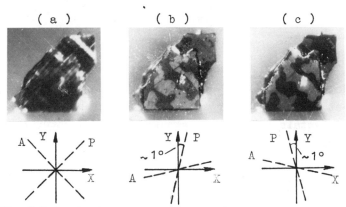

Fig.6 Polarized optical microscope photograph of $YBa_2Cu_3O_{7-\delta}$ sample. (a) Geometry of the experiment is the same as in Fig.1. (b) Polarizer is at 45° to one of two twin planes and analyzer is set at the extinction position for the region with the above twin domains. (c) The same geometry for the second twin plane. Contrast inversion of two families of microtwinned regions is well seen.

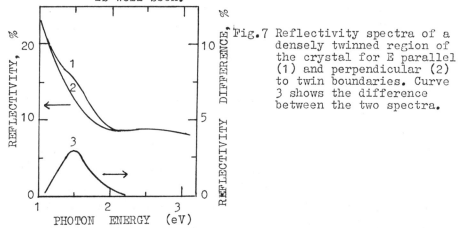

Fig.7 Reflectivity spectra of a densely twinned region of the crystal for E parallel (1) and perpendicular (2) to twin boundaries. Curve 3 shows the difference between the two spectra.

in the same geometry for single-domain regions, the reflectivity curve for E ∥ x' perpendicular to the boundaries is seen to be somewhat lower in the range 1-2 eV. This is likely to be due to electron states of twins localized at the boundaries, though contribution from size effects in the microtwinned region is also possible. In any case, one of the questions which any model should answer is why the maximum drop of reflectivity coincides with the reflectivity peak at 1.6 eV in the tetragonal phase.

RAMAN SPECTRA

Raman spectra of the $YBa_2Cu_3O_{7-\delta}$ compound are of great interest and intensively studied by different authors on both single crystals and ceramic samples /12,13/. However, as far as we know, spectra from single domains have not yet been obtained. Using the microRaman technique, we measured polarized spectra for single domains and also for microtwinned regions of $YBa_2Cu_3O_{7-\delta}$ crystals. The size of the examined region was 10 μm and the incident power was $\sim$ 1 mW. Spectra were excited by the 5145-A line of an Ar^+ laser, a backward scattering geometry was used.

The Raman spectrum from a single domain for the α_{xx} polarization is shown in Fig.8. No significant differences in spectra for $x \parallel a$ and $x \parallel b$ were found. The curve in Fig.8 is an averaged spectrum obtained from measurements on four crystals with domains 50-100 μm is size. The most intense band is at 335 cm^{-1} and there is also a number of less intense bands. The spectrum for α_{xy} ($x \parallel a$, $y \parallel b$) of a single domain does not exhibit any bands.

The band at 335 cm^{-1} was observed by a number of authors who studied spectra of $YBa_2Cu_3O_{7-\delta}$ and was assigned to bending vibration of oxygen atoms O2 and O3 in opposite phases /12/ (Ag symmetry). In addition to the vibration mentioned above, the crystal exhibits four vibrations of the Ag symmetry which are allowed in the α_{xx}-spectrum. However, as seen from Fig.8, there is a much greater number of bands in the spectrum. Therefore, the effects violating selection rules should be taken into account.

The content of oxygen in the crystals studied corresponds to $\delta \gtrsim 0.1$ and their translation symmetry is distorted. It is likely that a part of vibrations which appear in the α_{xx}-spectrum may be described by considering a finite chain formed by Cu1, O1, O4 atoms. An infinite chain consisting of such atoms has only one Ag vibration which is allowed in the α_{xx}-

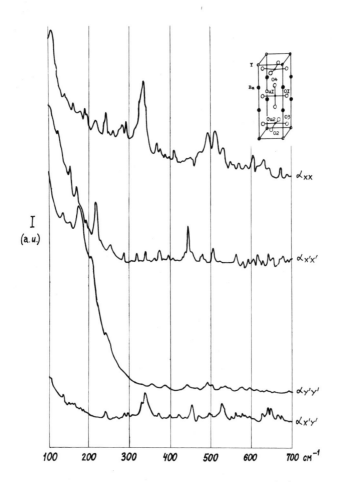

Fig.8 Raman spectra of a single domain (α_{xx}-polarization) and a microtwinned region ($\alpha_{x'x'}$, $\alpha_{y'y'}$, $\alpha_{x'y'}$-polarizations) for the YBa$_2$Cu$_3$O$_{7-\delta}$ crystal.

spectrum (stretching mode of O4 atoms in opposite phases at a frequency of ~ 500 cm^{-1} /12/). There are 2n+1 of Ag vibrations for a finite chain, where n is the number of O1 atoms. Specifically, the band corresponding to O4 vibration must split in this case into several bands which correspond to different phases of the vibration. Indeed, we observed a number of bands in the vicinity of 500 cm^{-1}, i.e., 470, 490, 510, and 530 cm^{-1}. A number of other bands may be interpreted in the same fashion.

Fairly pronounced specific features of the α_{xx}-spectrum attributable to vibrations of finite chains of copper and oxygen atoms indicate, in our opinion, that these chains have a regular length distibution. In other words, there is an ordered arrangement of oxygen vacancies. It is interesting to note that the model of the crystal lattice of $YBa_2Cu_3O_{6.87}$ /14/ including ordered oxygen vacancies is in a fairly good agreement with our data on the oxygen content in the crystals studied.

Fig.8 shows the Raman spectra of microtwinned regions for polarizations $\alpha_{x'x'}$, $\alpha_{y'y'}$, $\alpha_{x'y'}$ related to twin symmetry. It should be noted that the crystals have markedly different spectra and the figure gives the spectra for the sample which, in our opinion, demonstrates most clearly the basic effects typical of microtwinned regions. The $\alpha_{x'x'}$-spectrum has a rather large number of narrow lines, many of which correlate fairly well with frequencies of infrared and Raman active phonons of the $YBa_2Cu_3O_{7-\delta}$ crystal /13/. Specifically, in the neighbourhood of the Raman active mode at ~ 335 cm^{-1}, there is its IR active Davydov's pair at ~ 318 cm^{-1}. A similar doublet is seen in the vicinity of 500 cm^{-1} (506 cm^{-1}, 478 cm^{-1}).

Particular attention should be paid to the halfwidth of lines in the $\alpha_{x'x'}$-spectrum Fig.9 compares the shapes of the 335-cm^{-1} line in the α_{xx}- and $\alpha_{x'x'}$-spectra. The line is much more narrow in the $\alpha_{x'x'}$-spectrum. This effect is likely to be due to the fact that we observe the crystal's own phonon in the α_{xx}-spectrum and there is a phonon localized at the twin boundary in the $\alpha_{x'x'}$-spectrum.

The low-frequency modes are extremely active in the $\alpha_{x'x'}$-spectrum, especially the line at 174 cm^{-1}. These vibrations should be attributed to the motion of heavy Y.Ba,Cu atoms along the twin boundaries.

The spectra considered and also the $\alpha_{x'y'}$-spectrum

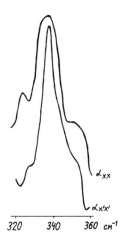

Fig.9 The 335-cm^{-1} line in the α_{xx}-spectrum of a single domain and in the $\alpha_{x'x'}$-spectrum of a microtwinned region.

show that microtwinning affects the Raman spectrum of the $YBa_2Cu_3O_{7-\delta}$ crystal. A high intensity of the modes localized at twin boundaries may be explained by a high concentration of boundary regions (30-50 Å in width /15,16/) and by an increasing Raman scattering crosssection due to transition between electron states at the twin boundaries at 1.5 eV (Fig.7). The fact that Raman spectra from microtwinned regions differ from one crystal to another is consistent with the assumption that microtwinning affects the spectra, because the structure of boundary regions may be different for different crystals and even for different regions of a crystal.

CONCLUSION

The optical studies of electron and phonon spectra of single domains and microtwinned regions of $YBa_2Cu_3O_{7-\delta}$ crystals have revealed a strong anisotropy of electron properties of single-crystal domains of $YBa_2Cu_3O_{7-\delta}$ in the a-b plane. The microtwinned regions have been observed to exhibit anisotropy of electron properties which has another origin and is attributable to twin boundaries.

It is shown that the Raman spectra of $YBa_2Cu_3O_{7-\delta}$ show not only the crystal's own phonons, but also vibrations of finite copper-oxygen chains and vibrations localized at the twin boundaries.

REFERENCES

1. Hidaka Y., Enomoto Y., Suzuki M., Oda M., and Moriwaki K., Jap.J.Appl.Phys., Part 2, 1987, v.26, L377.
2. Tozer S.W., Kleinsasser A.W., Penney T., Kaiser D., and Holtzberg F., Phys.Rev.Lett., 1987, v.59, p.1768.
3. Hikita M., Tajima Y., Katsui A., Hidaka Y., Iwata T., and Tsurumi S., Phys.Rev., 1987, v. B36, p.7199.
4. Suzuki M., Oda M., Enomoto Y., and Muzakami T., Jap.J.Appl.Phys., Part 2, 1987, v.26, L2052.
5. Wang X., Nanba T., Ikezawa W., Hayashi S., and Komatsu H., Jap.J.Appl.Phys., Part 2, v.26, 1987, L2023.
6. Petrov M.P., Grachev A.I., Krasinkova M.V., Nechitailov A.A., Prokofiev V.V., Poborchy V.V., Shagin S.I., and Kartenko N.F., Pis'ma v Zh.Tekh.Fiz., 1988, v.14, p.748.
7. Semba K., Suzuki H., Katsu A., Hidaka Y., Hikita M., and Tsurumi S., Jap.J.Appl.Phys., 1987, v.26, L1645.
8. Orenstein J., Thomas G.A., Rapkine D.H., Bethea C.G., Levine B.F., Cava R.J., Rietman E.A., and Jennson D.W., Phys. Rev., 1987, v.B36, p.729.
9. Kamaras K., Potter C.D., Doss M.G., Herr S.L., Tanner D.B., Bonn D.A., Greedan J.E., O'Reilly A.H., Stager C.V., and Timusk T., Phys.Rev.Lett., 1987, v.59, p.919.
10. Wang X., Nanba T., Ikezawa W., Isikawa Y., Mori K., Kobayshi K., Kasai K., Sato K., and Fukase T., Jap.J.Appl.Phys., 1987, v.26, L1391.
11. Verkin B.I., Dmitriev V.M., Dikin D.A., Kosmina M.B., Mitkevich V.V., Prikhodko O.R., Prokopovich S.F., Svetlov V.N., Seminogenko V.P., Khrostenko E.V., and Churilov G.E., Fiz.Nizk.Temp., 1988, v.14, p.218.
12. Krol D.M., Stalova M., Weber W., Schneemeyer L.F., Waszczak J.V., Zahurak S.M., Kosinski S.G., Phys.Rev., 1988, v.B36, p.8325.
13. Cardona M., Liu R., Thosen C., Bauer M., Genzel L., König W., Wittlin A., Amador U., Barahma M., Fernandez F., Otero C., Sáez R., Sol.St.Commun., 1988, v.65, p.71.
14. Hemly R.J., Mao H.K., Phys.Rev.Lett., 1987, v.58, p.2340.

15. Hirotsu Y.Y., Nakamura Y., Murata Y., Nagakaru S., Nishihara T., Takata M., Jap.J.Appl.Phys., 1987, v.26, L1168.
16. Sarikaya M., Kikuchi R., Aksay J.A., Physika, 1988, v. C152, p.161.

IR REFLECTIVITY SPECTRA OF $YBa_2Cu_3O_{7-x}$ SINGLE CRYSTALS

Bazhenov A.V., Gorbunov A.V., Timofeev V.B.

Institute of Solid State Physics, USSR Academy of Sciences, Chernogolovka, Moscow district, 142432, USSR

Abstract

Reflectivity spectra from the basal surface of tetragonal $YBa_2Cu_3O_{6+x}$ single crystals, containing orthorhombic phase inclusions, have been studied over a spectral range of 50-15000 cm^{-1}. The spectrum of planar optical phonons has been measured and the classification of the modes has been proposed. The electronic excitations have been observed at 0.6 and 1.5 eV. As a result of study of the temperature dependence of the reflectivity spectra and the effect of a low temperature anneal of the single crystals on the reflectivity spectrum a specific feature has been observed which is attributed to the manifestation of a superconducting gap $2\Delta = 3.2\ k\ T_c$.

Now high-T_c superconductors of $YBa_2Cu_3O_{7-x}$ (1-2-3) are being intensively studied by IR spectroscopy method. Despite numerous publications devoted, first of all, to investigation of IR reflectivity spectra R, the question concerning the nature of the spectrum of optical phonons, electronic excitations, the magnitude of the superconducting gap remains contraversial. This is due to the fact that ceramics or, at best, films were studied.

In analyzing R of these materials one encounters difficulties associated with necessity to take account of the effects of anisotropy and the processes of diffusive scattering of light. It does not make it possible to conduct correct measurements of polarization features of optical spectra, to employ Kramers-Kronig's analysis in order to reveal the spectral dependence of the dielectric permeability and the high frequency conductivity.

This work deals with the investigation of R of single crystals 1-2-3 in the range from 50 to 15000 cm^{-1}. The single crystals with T_c=60 K had the tetragonal structure with a considerable fraction of the orthorhombic phase, whose degree of orthorhombicity was by a factor of 1.5 smaller compared to the crystals with T_c=93 K. Under the transition of the samples to the superconducting state we have observed in R a specific feature, whose spectral characteristics altered at a low temperature anneal of the sample in a vacuum. This we attribute to the manifestation of the superconducting gap $2\Delta = (3.2\pm 0.1) k T_c$. In the tetragonal phase of 1-2-3 the spectrum of planar optical vibrations (in the plane perpendicular to the c-axis of the crystal) has been examined, the Kramers-Kronig analysis has been executed, the phonon classification has been proposed. The nature of two additional vibrational modes has been discussed. Finally, the electronic excitations have been observed at 0.6 and 1.5 eV at 300 K.

Experimental technique

Single crystals with T_c=60 K were grown from the solution in the melt with an excess of CuO. The transition to the superconducting state was detected from the temperature dependence of the magnetic susceptibility by means of Schielding and Meissner's effect. R was measured by a Fourier-spectrometer in the geometry close to the normal reflectivity of light with the wave vector $\bar{q} \parallel \bar{c}$ in the polarization $\bar{E} \perp \bar{c}$ (the developed surface of the samples was the basal (001) plane). The dimensions of the samples not exceeding 1.3x1.3x0.2 mm^3. In order to increase the signal-to-noise ratio, an oriented mosaic of 4-6 crystals was collected. To measure the temperature dependence of R in the range from 300 to 5 K the samples were fixed on a cold finger in a vacuum cavity of a continuous-flow helium cryostat. In order to obtain samples with preferentially tetragonal structure the samples were subjected to low temperature annealing in a vacuum cavity of the cryostat at 350 K for 8 hours.

An analysis of the phase composition and the real structure of annealed single crystals was executed by X-ray method. A detailed structural analysis was carried out by a high resolution angular scanning topography method /1,2/, the photographs were taken using a $Cu_{k\alpha}$ radiation on the (200) reflection.

Experimental results and discussion

1. <u>Reflectivity spectra in far IR region and their temperature dependence</u>. In as-grown crystals the reflectivity R (ν) in far IR region grows monotonically with diminishing wave number ν (spectra 1,2 in fig.1). Such a behaviour of R(ν) indicates a considerable contribution of free carriers to the dielectric permeability of the crystal. The spectrum exhibits a number of lines whose spectral position is markedly different from the optical phonons spectrum, observed in a superconducting ceramic with the orthorhombic structure (T_c=92 K) /3/. A decrease of the oxygen content in the single crystals as a result of low temperature anneal has led to a decreased contribution of free carriers to the dielectric permeability. The spectrum of the annealed single crystal is characterized by a low mean value of R(ν) and has the form peculiar for dielectrics (spectra 3,4 in fig.1). The spectrum consists of prominent lines which are naturally to be associated with the excitation of optical phonons. Here, the spectral position of the main spectral lines did not alter after the low temperature anneal.

We have attempted to detect the superconducting gap in the reflectivity spectra of the single crystals. The as-grown samples exhibited a slight increase of R(ν) in the region $\nu \leq 270$ cm^{-1} at $T \leq T_c$ (spectra 1,2 in fig.1a). This feature is clearly seen in difference spectra $(R_S-R_N) \cdot R_N^{-1}$ (R_S and R_N are reflectivity spectra in the superconducting and normal states of the sample, respectively), presented in fig. 2a (curves 1,2). Inasmuch as the reflectivity spectrum changes monotonically with varying the temperature from 300 to 60 K, we have measured the temperature dependence of the differential reflectivity spectra $\delta(\nu) = (R_T - R_{(T+10)}) \cdot R_{(T+10)}^{-1}$, the temperature step being 10 K. It appears that with $T > 60$ K $\delta(\nu)$ is practically independent

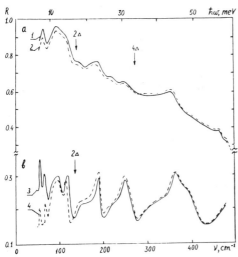

Fig.1. Reflectivity spectrum from the basal plane of $YBa_2Cu_3O_{6+x}$ single crystals with $T_c=60$ K (the wave vector of light $\bar{q} \parallel \bar{c}$, the electric vector $\bar{E} \perp \bar{C}$): a (curves 1 and 2) and b (curves 3 and 4) respectively, are spectra of as-grown and vacuum annealed single crystals. Spectra 1 and 3 are measured at T=5 K, 2 and 4 - at 70 K.

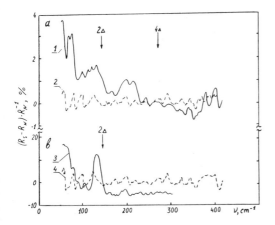

Fig.2. Difference reflectivity spectra of as-grown (a) and annealed (b) $YBa_2Cu_3O_{6+x}$ single crystals with $T_c=60$ K. R_N are measured at 70 K, R_S at 60 and 5 K, respectively, for spectra 2, 4 and 1, 2.

of ν (curve 2 in fig. 2a). However $\delta(\nu) \neq 0$, which introduces errors in the analysis of the difference reflectivity spectra $(R_S-R_N)R_N^{-1}$. It was shown in /4/ that the account taken of the temperature dependence of $\delta(\nu)$ is especially important in the case of inhomogeneous samples with a large fraction of the dielectric phase. So, in investigating the reflectivity spectra of figs 1 and 2 we have taken into account the monotonic ν-independent alteration of $\delta(\nu)$. At T=50 K $\delta(\nu)$ alters drastically like spectrum 1 in fig. 2a, at T 40 K $\delta(\nu)$, again, is independent of ν. So, the observed specific feature of the reflectivity spectra (fig. 1a) and of their difference spectra (fig. 2a) is formed at temperatures close to T_c=60 K, a maximal rate of the feature formation is observed at T=50 K. This suggests that this feature is the manifestation of the superconducting gap in the reflectivity spectrum.

At the same time these results do not allow one to determine the superconducting gap magnitude. It is known /5,6/ that in a perfect superconductor with the fraction of the superconducting phase f =1, $R_S(\nu)$=1 (100 %) at $\nu \leq 2\Delta$, and intersection of the spectra R_S and R_N is at ν_o=4Δ. The experimentally observable values $R_S(\nu) \leq 0.96$ (96 %) indicate that $f < 1$. Considering that the reflectivity of the superconducting phase is 1, then by means of the relationship $f + R_N (1-f) \cong R_S$ on the base of the spectra of fig. 1a, we obtain f =0.15. In accord with /5,6/, in the case of superconducting phase inclusions uniformly distributed over the dielectric matrix, the point of intersection of the difference spectra shifts towards the long-wave side with decreasing f, while the spectral position of maximum of superconducting-normal reflectivity difference depends comparatively weakly on f, remaining near ν =2Δ. With $f \ll 1$ a promonent feature in the absorption spectrum (and, accordingly, in the reflectivity spectrum) is observed at ν = (2÷2.5)Δ /7/. So, a variation of the superconducting phase content enables one, on the one hand, to conduct control experiments at detecting the superconducting gap by IR reflectivity method, on the other hand, to determine the magnitude of the superconducting gap in composites, consisting of a dielectric matrix and superconducting inclusions. In view of this, we have subjected the single crystals to a low temperature anneal in a vacuum. The magnetic susceptibility measurements showed that after the anneal the fraction of the superconducting phase reduced substantially to ~5 %. This value agrees with f = 4 %, obtained from the analysis of the reflectivity spectra. It turns out that, as a result of anneal the intersection of the normal and superconducting state spectra has shifted towards ν_o=150 cm^{-1} (spectra 3, 4 in figs 1 and 2) and near to ν_o a maximum is observed at ν =(132$\pm$3)cm^{-1}. Like in the as-grown crystals this feature is developing at temperatures close to T_c=60 K, a maximal formation rate is at 50 K.

In such a case the above results may be explained as follows. In the annealed single crystals the superconducting phase in the form of microinclusions is uniformly distributed over the dielectric matrix. The spectral position of the high-energy maximum of the difference spectra (spectrum 3 fig.2) occurring at ν =(132$\pm$3) cm^{-1} and the point of the spectra intersection ν_o=(150+3) cm^{-1} are close to each other. The 2Δ value is probably between them and is close to the value 2Δ=(140$\pm$5)cm^{-1}. In unannealed single crystals, containing a larger fraction of the superconducting phase the latter is distributed nonuniformly as macroscopic formations. Therefore ν_o=(270$\pm$15) cm^{-1} in these crystals is close to the 4Δ value. Accordingly, the value 2Δ=(135$\pm$7) cm^{-1} is practically coincident with the spectral position of the high energy maximum of the difference spectra of an as-grown sample (spectrum 1, fig. 2a). Undoubtedly, the absence of 100 % reflectivity at $\nu \leq 2\Delta$ decreases the validity of the experimental data processing in terms of the approximation /5,6/. Nevertheless, allowing for the afore arguments, we believe that in this experiment the superconducting gap has manifested itself. So, in the basal plane of 1-2-3 single crystal with T_c=60 K 2Δ=(140$\pm$4) cm^{-1}=(3.2$\pm$0.1) k T_c.

The difference spectra of as-grown and annealed samples (fig. 2) exhibit the structure that partly correlates with the spectral position of phonons. An analogous effect was earlier observed in superconducting ceramics with T_c=90 K /3,4,6/. This structure is, seemingly, determined by a change of the phonon lifetime at $T<T_c$, since at $T<T_c$ scattering on normal electrons is absent in superconducting phase. Besides this structure one observes an increase of (R_S-R_N) with decreasing ν below 80 cm^{-1}, whereas in accord with the theoretical representations of /5-7/ (R_S-R_N) must monotonically tend to zero as ν decreases from 2Δ to 0. In our opinion, this difference results

from the dependence of the free carriers lifetime on frequency in the region of low frequencies at $T > T_c$.

In conclusion we shall consider the question concerning the nature of the superconducting phase in the investigated single crystals. As a result of X-ray measurements of the annealed single crystals /2/ angular scanning topographs were obtained with five crystal images: central reflection (most intensive) corresponded to the large bulk content of the tetragonal phase, four weak spots, surrounding it, corresponded to the orientational states of the orthorhombic phase. A relative content of the orthorhombic phase, estimated from the photometering, made up not more than several percents, which correlates with the estimations of the fraction of the superconducting phase, using the Meissner effect and the reflectivity spectra analysis. This suggests that in our case the superconducting phase is inclusions of the orthorhombic structure dissolved in the dielectric tetragonal matrix. An analysis of the topographs by means of an optical microscope has shown that the orthorhombic phase is uniformly distributed over the crystal, as was supposed for the annealed sample. Here, the degree of orthorhombicity of this phase appears by approximately 1.5 times smaller than in crystals with T_c=93 K (the lattice parameters ratio of the 60° phase, b/a =1.011 against 1.017 in the 93° phase) which agrees with the results of X-ray studies of 1-2-3 crystals with large oxygen deficiency (x=0.5) /8/.

2. <u>The optical-phonon spectrum</u>. The X-ray studies have shown that the dominant fraction of the investigated single crystals is tetragonal, therefore the theoretical group analysis of normal vibrations we considered for the tetragonal structure of $YBa_2Cu_3O_6$ (space group D_{4h}^1, P4/mmm /9/). A unit cell of this crystal contains 12 atoms, therefore the spectrum of normal vibrations consists of 36 modes: 33 optical and 3 acoustic. Among the optical vibrations there are Raman-active four total-symmetrical modes A_{1g}, one mode B_{1g} and five two-fold degenerated modes of E_g symmetry. Since the crystal is centrosymmetric, these modes must be inactive in the IR spectrum. In the tetragonal structure the mode B_{2u} is also inactive (it corresponds

to anti-phase vibrations of oxygen atoms O(2) and O(3) in the c-direction. For simplicity, we retain the conventional designations of atoms of the orthorhombic structure /10/ in describing the tetragonal phase). The IR spectra may exhibit 5 modes of the A_{2u} symmetry, corresponding to the motion of atoms along the $\bar{c}$ direction and 6 two-fold degenerated modes of the E_u symmetry, corresponding to planar atomic vibrations in the $\bar{a}\bar{b}$ plane. These results coincide with the analogous analysis, carried out in the works /10,11/. We have examined the reflectivity spectrum of the single crystal under a normal illumination of the basal plane (wave vector $\bar{q} \parallel \bar{c}$, electric vector $\bar{E} \perp \bar{c}$), therefore it is precisely the latter six out of the aforementioned modes that must be observed in the spectrum, presented in fig. 3. The spectra of the imaginary part of the dielectric permeability $Im\mathcal{E}$ and $-Im\mathcal{E}^{-1}$ (fig.3) whose maxima determine the spectral positions of TO and LO phonons, respectively, are obtained by the Kramers-Kronig transformation of the reflectivity spectra.

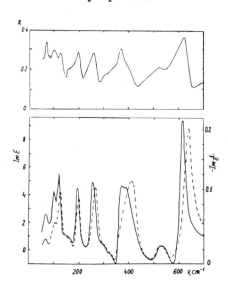

Fig.3. Optical phonons in the reflectivity spectrum $R(\nu)$ of tetragonal $YBa_2Cu_3O_{6+x}$ single crystals and the corresponding spectra of the imaginary part of the dielectric permeability $Im\mathcal{E}$ (solid curve) and $-Im\mathcal{E}^{-1}$ (dashed curve), whose maxima determine the spectral position of TO and LO phonons, respectively, at T=5 K.

The theoretical calculation of the normal vibration frequencies of $YBa_2Cu_3O_6$ in the shell model approximation was performed in /10/.

The results of this calculation for the case of planar vibrations are presented in fig.4. Table 1 presents the assignment of the frequencies of the main lines of the experimentally observable TO(LO)-vibrations (fig.3) with closest in the value calculated frequencies of the normal vibrations of fig.4 /10/. Apparently, in this assignment the connection of the

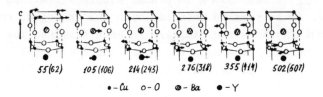

Fig.4. Schematic presentation of E_u symmetry normal vibrations in the tetragonal $YBa_2Cu_3O_{6+x}$ and the corresponding calculated values of the TO(LO)-vibration frequencies /10/.

Table 1. Experimental values of the frequencies of TO- and LO-phonons in the tetragonal single crystal of $YBa_2Cu_3O_{6+x}$ ($\bar{q} \parallel \bar{c}$, $\bar{E} \perp \bar{c}$). Comparison with the theoretical calculation /10/.

Mode	$\nu_{TO}(\nu_{LO})$, cm^{-1}		$(\nu_{LO}-\nu_{TO})$, cm^{-1}	
	Theory	Experim.	Theory	Experim.
CuI	55(62)	100(102)	7	2
Ba	105(106)	121(125)	1	4
Y	214(243)	191(196)	11	5
Cu2-02-03	276(318)	256(261)	42	5
04	355(414)	388(410)	59	22
Cu2-02-03	502(507)	613(637)	5	20
	-	530(530)	-	0
	-	68(68)	-	0

191(196) cm^{-1} lines with the planar vibration of yttrium is doubtless. Besides the close values of the calculated and experimentally observed frequencies, this assignment is supported by the fact that this line practically does not change its spectral position when transiting from the orthorhombic /3/ to the tetragonal phase. It is precisely this behaviour of the yttrium vibrations that may be expected basing on the fact that yttrium atoms are far away from the CuI-O1 plane, from which oxygen is liberated during the transformation to the tetragonal phase. In fact, the structural studies have shown that the environment of yttrium changes insignificantly at this transformation /12/. A satisfactory agreement between the experimentally observed and the culculated frequencies of the remaining modes supports the assignment, presented in Table 1. It should, however, be noted that the experimentally observed values ($\nu_{LO} - \nu_{TO}$) are, nevertheless, mush smaller than the calculated ones. Since ($\nu_{LO}^2 - \nu_{TO}^2$)~$Q^2 \cdot m^{-1}$ /13/, where Q and m are, respectively, the effective charges and masses of the vibrating atoms, the degree of the crystal ionicity appears smaller than that proposed in the calculations /10/ (O^{-2}, Ba^{+2}, Y^{+3}, $Cu^{1.66}$). A decrease of the effective charges of atoms may lead to a change of the modes assignment, presented in Table 1. The strongest disagreements between the experimental and theoretical values ($\nu_{LO} - \nu_{TO}$) are observed for the bending 256(261) cm^{-1} and stretching 613(637) cm^{-1} vibrations of Cu2-O2-O3 atoms. In accord with the calculation /10/ TO-LO splitting of the bending vibrations is greater than that of the stretching vibrations. Allowing for the fact that the experimentally observable TO-LO splitting of the stretching vibrations makes up 20 cm^{-1}, one can suppose, that the bending vibration of Cu2-O2-O3 corresponds to $\nu_{TO}(\nu_{LO})$=388(410) cm^{-1}. In this case, the bending vibration of O4 will be met by the frequency $\nu_{TO}(\nu_{LO})$=256(261) cm^{-1}. This assumption requires that additional theoretical calculations and experiments be carried out.

In conclusion we shall note that the reflectivity spectrum of tetragonal single crystal (fig.3) contains a greater number of lines than it was predicted by the theoretical group analysis, namely, 8 instead of the expected 6 lines. The "extra" 68 cm^{-1} line is different from the rest in that it does not change as the temperature is decreased from 300 down to 100 K, but at $T \leqslant 100$ K it transforms to two narrow lines 68 and 59 cm^{-1} (see fig.1). The "extra" 530 cm^{-1} line is distinguished from the rest of the lines by the absence of the TO-LO splitting. One of the possible reasons for the appearance of this line is excitation of local vibrations associated with the presence of some quantity of oxygen atoms O1 in the tetragonal structure of $YBa_2Cu_3O_{6+x}$. It is known that the tetragonal structure is realized at $0.3 \geqslant x \geqslant 0$ /14/. Here, the oxygen O1 is, seemingly, distributed over the crystal in a random manner without forming Cu1-O1 chains.

3. <u>Electronic transitions</u>. As a result of investigation of the reflectivity spectrum from the basal plane of a vacuum annealed single crystals in a spectral range from 50 to 15000 cm^{-1} (fig.5), we have observed two broad lines with maxima at 2500 and 13600 cm^{-1}. These lines are prominent at T=100 K. With increasing temperature up to 300 K the maxima are shifted towards the long-wave side, respectively, to 1500 and 13200 cm^{-1}. In doing so, both lines broaden and become not so prominent, the reflectance in the 2500 cm^{-1} maximum decreases significantly in contrast tp the reflectivity spectrum lines in the phonon excitation region ($\nu < 700$ cm^{-1}). Using the Kramers-Kronig transformation the reflectivity spectrum at 100 K was transformed to the absorption spectrum, shown in fig. 5 by curve 3. The 2500 and 13600 cm^{-1} lines are met by absorption spectrum maxima at 0.7 and 1.6 eV with absorptivity $\alpha \cong 5 \cdot 10^4$ cm^{-1}. With increasing the temperature up to 300 K, the absorption spectrum somewhat flattens and the maxima of the lines shift to 0.6 and 1.5 eV, respectively.

With account taken of the spectral position and the shape of these lines, it is natural to attribute them to electronic transitions. The analogous lines were observed in the absorption spectra of orthorhombic films of $YBa_2Cu_3O_{7-x}$ (x=0) with T_c=87-90 K /15/. The authors attributed these lines to d-d transitions of Cu^{++}. We assume that in the case of our tetragonal single crystals we also observe the optical transitions between the d-states of divalent copper ions (these states are split in the ligands crystalline field of the corresponding symmetry). The known studies of electron loss spectra and of photoemission spectra indicate that copper ions in yttrium-barium cuprates are divalent and that holes, arising in CuO_2-planes as a result of doping are mainly located in the oxygen sites /16-18/. The Cu^{++} ion corresponds to the electronic con-

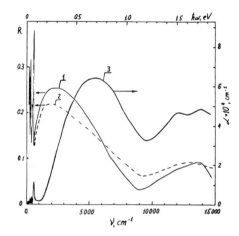

Fig.5. Reflectivity spectrum $R(\nu)$ of tetragonal $YBa_2Cu_3O_{6+x}$ single crystals ($\bar{q} \parallel \bar{c}$, $\bar{E} \perp \bar{C}$) at T=100 K (solid curve 1) and T=300 K (dashed curve 2). Spectrum 3 is absorption spectrum $\alpha(\nu)$, obtained by the Kramers-Kronig transformation of reflectivity spectrum 1.

figuration d^9, here, one hole arises in the d-shell. In the cubic symmetry field the five-fold degenerated d-state of the hole must split into two-fold degenerated e_g- and three-fold degenerated t_{2g}-states. In tetragonal and orthorhombic crystals there occurs a further symmetry decrease followed by additional splitting of the afore states. It is supposed that the d-hole has the $d(e_g)$ symmetry, corresponding to the electronic configuration $d(x^2-y^2)$ /12/. In view of this in the case of yttrium-barium cuprate of the tetragonal symmetry one expects three d-d-transitions, associated with the Cu^{++} ions /20,21/. The lowest transition is associated with excitations within e_g-orbitals. In our case, this corresponds to ~0.2 eV peak at 300 K (0.3 eV at 100 K), observable in the reflectivity spectra of tetragonal single crystals (fig.5). In the absorption spectrum of fig. 5 this transition is met by a 0.7 eV peak at 100 K (0.6 eV at 300 K). The existence of said electronic excitations and their direct experimental observation is of course of interest, since these excitations may be mediator of pairing of carriers (holes) in cuprate planes.

The authors are grateful to A.I.Shalynin and V.A.Larkin for the measurement of the magnetic susceptibility in the investigated samples.

References

1. Ossipyan Yu.A., Aphonikova N.S., Emel'chenko G.A., Parsamyan T.K., Shmyt'ko I.M., Shekhtman V.S. Pis'ma ZhETF, 1987, v.46, p.189.
2. Bazhenov A.V., Gorbunov A.V., Emel'chenko G.A., Parsamyan T.K., Tatarchenko V.A., Timofeev V.B. All-Union Seminar "Spectroscopy of High-T_c Superconductors". Moscow. 24 November 1987.
3. Bazhenov A.V., Gorbunov A.V., Klassen N.V., Kondakov S.F., Kukushkin I.V., Kulakovskii V.D., Misochko O.V., Timofeev V.B., Chernyshova L.I., Shepel B.N. "Novel superconductivity" Plenum Press, New York and London. 1987, p.893.
4. Bazhenov A.V., Gorbunov A.V. "Problems of High-T_c Superconductivity", p.2. Sverdlovsk. 1987, p.9.
5. Mattis D.C., Bardeen J. Phys.Rev. 1958, v.111, p.412.
6. Thomas G.A., Ng N.K., Millis A.J., Bhatt R.N., Cava R.S., Rietman E.A., Johnson D.W., Espinosa G.P., Vandenberg J.M. Phys.Rev.B, 1987, v.36, p.846.
7. Curtin W.A., Aschcroft N.W. Phys.Rev.B, 1985, v.31, p.3287.
8. Manthiram A., Swinnea J.S., Sni Z.T., Steinfink H., Goodenough J.B. J.Am.Chem.Soc., 1987, v.109, p.6667.
9. Murphy D.W., Sunshine S.A., Gallagher P.K., O'Bryan H.M., Cava R.J., Batlogg B., Dover R.B., Schneemeyer I.F., Zuhurak S.M. Proc.ACS Symp. "Chem. HT$_c$ Supercond." 1987, p.181.
10. Thomsen C., Cardona M., Kress W., Liu R., Genzel L., Baner M., Schönherr E. Sol.St.Comm., 1988, v.65, p.1139.
11. Burns G., Dacol F.H., Freitas P., Plaskett T.S., König W. Sol.St.Comm., 1987, v.64, p.471.
12. Santoro A., Miraglla S., Beech F., Sunshine S.A., Murphy D.W., Schneemeyer L.F., Waszczak J.V. Mat.Res.Bull., 1987. v.22, p.1007.

13. Burns G. Solid State Physics. Academic, New York, 1985, chapter 13.
14. Cava R.J., Batlogg B., Chen C.H., Rietman E.A., Zahnrak S.M., Werder D. Phys.Rev.B, 1987, v.36, p.5719.
15. Geserich H.P., Scheiber G., Geerk J., Li H.C., Linker G., Assmus W., Weber W. Europhys.Lett., 1988, v.6, p.277.
16. Fujimori A., Takayama-Muromachi E., Uchida Y., Okai B. Phys.Rev.B, 1987, v.35, p.8814.
17. Bianconi A., Castellano A.C., Sanotis M., Delogu P., Gargano A., Giorgi R. Sol.St.Comm., 1987, v.63, p.1135.
18. Rietschel H., Fink J., Gering E., Gompf F., Nocker N., Pintschovins L., Renker B., Reichardt W., Schmidt H., Weber W. Physica C, 1988, v.153-155, p.1067.
19. Mattheiss L.F., Hamann D.R. Sol.St.Comm., 1987, v.63, p.395.
20. Weber W. Z.Phys.B, to be published.
21. Geserich H.P., Koch B., Scheiber G., Geerk J., Li H.C., Linker G., Weber W., Assmus W. Physica C, 1988, v.153-155, p.661.

EFFECT OF OXYGEN CONCENTRATION
ON TRANSPORT PROPERTIES OF $YBa_2Cu_3O_x$, (6<x<7).

A.V. Samoilov, A.A. Yurgens, N.V. Zavaritsky

(Institute for Physical Problems, Academy of Sciences,
Moscow, USSR)

We have measured the electrical resistivity ρ, thermal conductivity κ and thermoelectric power α of a series of high temperature superconducting compounds of the form $YBa_2Cu_3O_x$ with 6<x<7 in temperature range 4.2-260 K and the pressure dependence of α and ρ at room temperature.

It is found the transport properties of ceramic samples with oxygen content x=7 are nearly the same as published elsewhere [1,2], but they are changed essentially as x alters from 7.

Using the electrical resistivity data and the Wiedemann-Franz low, we estimate a magnitude of the electronic component of the thermal conductivity to be an order of magnitude smaller then the measured thermal conductivity and conclude that approximatelly all of the heat is conducted by lattice. It is experimentally shown the thermal conductivity of samples with x=6 to be larger than κ of others with x>6. This fact may be explained by decreasing of the phonon scattering due to the nearly total absence of electrons and oxygen-placing defects in the compound with x=6. From comparison of thermal conductivities of samples with x=6 and x=7 at 100 K one can obtain an estimation of phonon-electron relaxation time $\tau_{ph,e}$ $\sim 10^{-12}$ s, which is typical for ordinary superconductors (see, for example, tin [3]).

The measured dependence of ρ and α on a pressure P<12 kbar is approximately linear. The absolute value of derivative $\frac{\partial \ln(\rho)}{\partial P}$ increases but the value of derivative $\frac{\partial \ln(\alpha)}{\partial P}$ goes to zero as oxygen content alters from 7 to 6.4.

INTRODUCTION.

Since the discovery of high-T_c superconductivity in $YBa_2Cu_3O_x$ [4], many studies have been made, concerning different aspects of the problem of explaining such high T_c. Central to this question is whether these materials are standard Bardeen-Cooper-Shrieffer (BCS)-type superconductors, i.e. whether the electron-phonon interaction is large enough to support the high transition temperatures observed. Some investigations, such as absence of a copper [5] or barium [6] isotope effect and the absence of saturation of the high- temperature resistivity [7] suggest the existence of an electron coupling mechanism different from BCS one. Recent observations of a small oxygen isotope effect [8], on the other hand, raise one's hopes of the BCS- type explanation for the high-T_c superconductivity. Thus one conclude the problem of the mechanism of high-T_c superconductivity in 1-2-3 system remains unsolved.

Transport properties: the electrical resistivity ρ, thermal conductivity κ and thermoelectric power α make it possible to obtain very usefull information about interactions between carriers and phonons. However, the thermoelectric power of 1-2-3 compounds is too sensitive to the methods of a sample preparation [9] to be reliable interpreted in terms of the electron-phonon interaction.

Using the resistivity and thermal conductivity data in normal and superconducting states, one can estimate the strength of the electron-phonon coupling. There are significant differences between such estimations. The weak coupling estimation [7] leads to a conclusion of non-phonon origin of the high-T_c superconductivity, but the strong one [10,11] is well enough to explain high T_c by the BCS-type theory.

In addition, the oxygen content x may serve as a probe to verify any conclusion about phonon-electron interactions. Now it is well known the critical temperature [12] and carrier density [13] go to zero as oxygen content diminishes to 6. The thermal conductivity of $YBa_2Cu_3O_x$ compounds at a temperature about 100 K is predominatingly due to the lattice: the electronic component estimated is at most 10-15% of the total thermal conductivity.

Since the phonons can be intensively scattered by the electrons, one can expect the enchancement of κ with diminishing of x. The shot-wave phonons, typical for temperatures $T>\Theta/4\sim100$ K (Θ - the Debye temperature), are scattered by the oxygen vacancies and other crystal defects also. The more oxygen atoms are removed the more vacancies are produced.

So, the thermal conductivity is influenced by the oxygen content x by two opposite ways. One of them, the decreasing of the elecrton's density results in the increasing of κ, but another, the appearance of the oxygen vacancies leads to the decreasing of κ. According to this point of view, the sample with $x=6$ is to show the biggest thermal conductivity among the rest samples with $x>6$ because of the absence of any free electrons and oxygen-placing defects in it.

Taking into account this two aspects and using our experimental thermal conductivity data, we estimate the scattering time $\tau_{ph,e}$ to be 10^{-12} s at 100 K, which is comparable with that of ordinary superconductors [3].

The density of the electrons in high-T_c compounds is much smaller than that in the metals. So, having the same relaxation time $\tau_{ph,e}$, the high-T_c compound is to show the phonon-electron coupling stronger than that in the common superconductors.

EXPERIMENT.

The initial compound $YBa_2Cu_3O_x$ with nearly stoichiometric oxygen content $x=7$ were prepared according to the standard recipes: the Y, Cu, oxides and Ba carbonate were mixed in the appropriate proportions, pressed into the pellets and then were annealed at the temperature 900-980 °C within a week in an air atmosphere. The pellets were then ground into a powder. The samples were fabricated by pressing the powder obtained into the parallelepipeds 1.5x2.5x20 mm^3 with aid of the special press-form. So prepared parallelepipeds were annealed again in the oxygen gas flow for 9-31 h. at the temperature 940-980 °C to sinter grains. After that the temperature were slowly diminished down to 500-600 °C. The samples were exposed at this temperature for 6-10 h. and then were slowly (within 12 h.) cooled down to the room temperature. The X-ray analysis of the samples thus obtained shows a perfect orthorhombic-phase structure. The values of the a,b,c axis indicate the nearly stoichiometric oxygen content [14]. Oxygen removal was accomplished by annealing the sample under a vacuum 10^{-2}-10^{-4} torr at t>400 °C for several hours. The re-intercolation of the oxygen was done by reannealing the sample at 500-600 °C in the oxygen gas flow. The change of the oxygen content x were measured by weighting the samples before and after annealing or/and by condensing the evolving oxygen into a liquid helium sorber and measuring the gas pressure as the sorber was re-heated up to a room temperature after the annealing prosedure.

According to the dependence of the axis magnitude and a/b ratio on the oxygen content [14] we have examined the uniformity of the oxygen distribution after the removal process across the sample section. The X-ray analysis of the outer and inner parts of a probe sample has shown the acceptable uniformity: the difference between oxygen contents was less than 0.1.

All transport properties were measured in the vacuum container, placed into a liquid helium bath at the constant electrical and thermal currents by the standard method. Leads were attached to a sample using the silver paint and the manganin-constantan thermocouple was used to monitor the temperature and temperature gradient. The temperature dependence of the thermoelecric power of such thermocouple were determined in the additional experiments with aid of the platinum thermometer. The potential wires were also made from the manganin one. All wires were as thin (30 μm) as possible to prevent the heat losses. To prevent the energy losses by the radiation as temperature was rised up to 100-200 K the sample was surrounded by the copper shield thermally attached to the lower temperature edge of the specimen.

Two heaters were used to produce the temperature gradient (~1.5-5 K) along the sample at 100 K and to increase the average temperature of sample up to 300 K. The corresponding powers were 10 mWt and 0.2 Wt maximum.

The pressure measurements at a room temperature were made using the high pressure chamber [15] in which the hydrostatic pressures up to 12-15 kbar could be easily obtained.

All electical potentials were measured by the dc-potentiometer with sensitivity 10^{-8} V. We estimate our measurements to have an accuracy not worse than 5%. One can find some information about the samples under study in the Table.

N	T_c	ΔT_c	x	preparation
1/0	92	1	7.0	common method
1/1	--	--	6.0	annealing under vacuum for 48 h. at 700°C
2	90	1	7.0	common method
3/0	92	1	7.0	common method
3/1	88	3	6.75	annealing under vacuum for 6 h. at 420°C
4/0	92	1	7.0	common method
4/1	71	15	6.7	annealing under vacuum for 6 h. at 435°C
4/2	56	12	6.53	annealing under vacuum for 6 h. at 455°C
4/3	--	--	6.31	annealing under vacuum for 6 h. at 470°C
4/4	--	--	6.25	annealing under vacuum for 6 h. at 625°C
4/5	91	1	6.77	annealing in oxygen for 8 h. at 540°C
4/6	--	--	6.22	annealing under vacuum for 6 h. at 450°C
4/7	--	--	6.0	annealing under vacuum for 40 h. at 675°C

ΔT_c were determined as the difference between the temperatures, corresponding to 90% and 10% of the resistivity transition curve.

Note that the different interges under the slash in a specimen number mean the consistent annealing prosedures with one and the same sample. I.e. the initial sample number 4/0 is firstly annealed at 435 °C for 6 h. - this is the sample number 4/1; it is investigated and then is annealed again at 455 C for 6 h. 455 °C C for 6 h. - this is the sample 4/2 now, and so on.

RESULTS AND DISCUSSION.
-- Resistivity.

Figure 1 exhibits the resistivity of samples with the different oxygen content vs. temperature and the dependence of resistivity upon the oxygen content at the temperature T=120K.

The large increase of the resistivity with diminishing of x is due to the decrease of the carrier density [13]. The peak near x=6.5 appears probably because of the orthorhombic-to-tetragonal

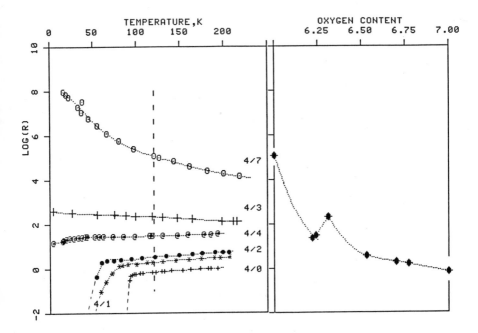

Figure 1. $\log(\rho)$ (ρ in $m\Omega \cdot cm$) vs. temperature (left) and $\log(\rho)$ vs. oxygen content x at the temperature 120 K (right).

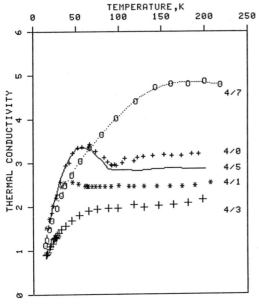

Figure 2. The thermal conductivity (in units $10^{-2}\ Wt \cdot cm^{-1} \cdot K^{-1}$) vs. temperature for different samples.

transition taking place in the 1-2-3 compounds with non-stoichiometric oxygen content. This peak was carefully investigated in [16].
-- Thermal conductivity.

In Figure 2 the thermal conductivities of the samples N4/0-4/7 are plotted as a function of temperature. The thermal conductivities of our samples are predominatingly due to the phonons, we estimate the electronic part to be not larger than 10-15% of the total thermal conductivity. We have done such estimation using the Wiedemann-Franz low and the smallest resistivty value obtained (0.5 mΩ·cm) at a temperature just above the transition temperature.

In Figure 2 one can see the thermal conductivity behavour of the initial sample N4/0 is analogues to that of the ordinary superconductors contaminated by impurities [17] in which the general part of the thermal conductivity is due to the phonons, scattered mainly by the electrons. The enhancement of κ just below T_c is explained by the exponential reduction of the electronic exitations, which scatter phonons, as electrons are condensed into the Cooper-pair Bose liquid [18]. The temperature at which the $\kappa(T)$-curve has a break (for all superconducting samples) corresponds to their resistivity transition temperature. The superconducting transition temperatures of samples 4/3, 4/4, 4/6, 4/7 is lower than our experimental temperature range (4-260 K).

According to this point of view the less is the number of electrons, the less is the scattering of phonons. Thus one expect the magnitude of the κ to increase and difference between thermal conductivities in normal and superconducting states to vanish as the carrier density falls. We actually have found the decreasing of maximum in the κ vs. T dependence in the superconducting state with diminishing of x. The temperature dependence of the thermal conductivity of the sample 4/2 has a little maximum, elevated over a normal state thermal conductivity, but the $\kappa(T)$ of the sample 4/4 already shows the absence of any extremums.

Reannealed in oxygen gas flow the sample has the same thermal conductivity as the initial one (see the curve, corresponded to the sample N4/5). Reversibility of the removal-re-intercolating process indicates the change of the thermal conductivity is due not to enlarging of the porosity but mainly to the change in κ of the general material.

In Figure 2 one can see the decreasing of κ with decreasing of x at any temperature above T_c. This fact is inconsistant with all above. One is to conclude either electronic part of the thermal conductivity is larger than estimated and the phonons are not dominant in a heat transfer, or the phonon scattering process is influenced not only by the electron-system changes but something else. We suppose the second suggestion is more acceptable.

Actually, any crystal imperfections, such as oxygen vacancies accidentally distributed in a lattice from a cell to cell can strongly scatter the short-wave phonons, typical for the temperatures $T>\theta/4 \sim 100$ K and so can produce additional thermal resistance, influenced by the oxygen deficiency. The more oxygen atoms is removed the more scattering centers is appeared. Extracting more and more oxygen atoms from the lattice one finally obtain the tetragonal insulating compound $YBa_2Cu_3O_6$ with also stoichiometric oxygen content. To the contrary with $YBa_2Cu_3O_{7-\delta}$ in this compound the absence of additional oxygen atoms means the absence of imperfections. Thus one may expect the increasing of κ if x is close enough to 6. Really, in Figure 2 one can find the curve, corresponding to the sample with $x=6$ to have thermal conductivity larger than that of other's from the series n4 nearly twice. Similar increasing of $\kappa(x; T>T_c)$ was observed for the sample n1/1

One can roughly write the thermal conductivity of lattice in the form:

$$\kappa \sim C \cdot s^2 \cdot \tau$$

$$1/\tau = 1/\tau_{ph,b} + 1/\tau_{ph,d} + 1/\tau_{ph,ph} + 1/\tau_{ph,e} ,$$

there C is the heat capacity per unit volume, s - is the sound velosity, and the τ with different subscripts - are the relaxation times due to the phonon scattering by the boundaries, defects, phonons and electrons respectively.

Let us now compare two extreme stoichiometric compounds: $YBa_2Cu_3O_7$ and $YBa_2Cu_3O_6$. The heat capacity of both compounds are nearly the same [20], the sound velocity increases by ~9% only as x jumps from 7 to 6 [21]. So, the enchancement of the thermal

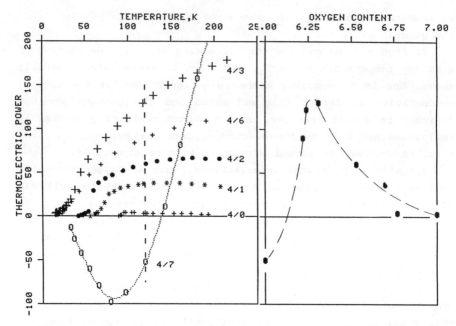

Figure 3. The thermoelectric power α (μV/K) vs. temperature (left) and α(T=120 K; x)-dependence (right).

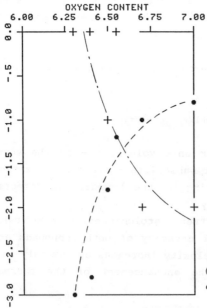

Figure 4. The relative derivatives $\frac{\partial \ln(\rho)}{\partial P}$ (points) and $\frac{\partial \ln(\alpha)}{\partial P}$ (crosses) (in units %·kbar^{-1}) vs. oxygen content.

conductivity of one compound as compared with another is due mainly to vanishing of carriers. From the comparison of these two thermal conductivities we have estimated $\tau_{ph,e}$ to be about 10^{-12} s at 100 K, which is comparable with $\tau_{ph,e}$ of ordinary superconductors at the same temperature [3].

The density of the carriers in common metals is much larger than that in the high-T_c materials. Thus one conclude the electron-phonon coupling is to be stronger than that in the ordinary superconductors.

We don't give any speculative estimations of the electron-phonon coupling value or any final sentence about the mechanism of the high-T_c superconductivity here. There are many of them, depending mainly on the author's faith in BCS or any other theory.

One additional aspect, connected with the oxygen deficiency araises from the intented view on the problem. There are several works discussing the existence or absence of a plato of the T_c vs. x dependence [12,21]. There is a suggestion about the order-disorder transition in the oxygen vacancies distribution, taking place in 1-2-3 system. The question is: what places in a lattice the oxygen atoms leave, and how does the method of a sample preparation influence upon the leaving process. We believe the thermal conductivity may serve as a probe characteristic to examine the character of the oxygen distribution. The larger is κ, the more ordered is the distribution of the oxygen vacancies at the same total content x. To prove this proposal it is necessary to produce the samples with higher uniformity of oxygen content across and along the specimens and to provide more accurate measurements of x we have done.

-- Thermoelectric power.

The thermoelectric power of several samples vs. temperature is presented in Figure 3. The dependence $\alpha(T)$ for the stoichiometric compound is similar in many works [22,23]. Authors [22] suggest the total thermoelectric power α is the sum of the diffusive part and phonon drag component: $\alpha=\alpha_{diff}+\alpha_{drag}$. It was shown, that $\alpha_{diff}\sim T$ and $\alpha_{drag}\sim 1/T$. The theory [24] explains experimental behavour of $\alpha(T)$ by a thin peak in the density of states near the fermi-level, ignoring the electron-phonon interaction.

The absolute value of the thermoelectric power of $YBa_2Cu_3O_7$ in the normal state is as small (2-10 $\mu V/K$) as that of ordinary metals and a positive sign of α indicates that the electric current is carried by holes. However, there is an investigation of nearly stoichiometric $YBa_2Cu_3O_{7-\delta}$, which shows an opposite sign of α due to conditions of the preparation [23]. The authors [23] suppose different signes appear from the different zones contributing the thermoelectric power.

The dependence of α on x at the temperature T=120 K is plotted in Figure 3. Large enchancement of α may be explained by two ways. Firstly, it may be explained by the structural transition from the orthorhombic to tetragonal lattice at the oxygen content x=6.4, and secondly, by the Mott transition due to a small carrier density and disorder in the oxygen' vacancies distribution. The symmetry transition may be accomplished by the appearance or disappearance of any small parts of the Fermy surface (the Lifshitz transition [25]). Such changes in the electronic system results in the most essential changes of the thermoelectric power due to the singularity in the energy dependence of the electron mean free parth [26]: an electron may get to the small parts of the Fermi surface with a group velosity equal nearly zero by scattering on the defects. Thus a new way of the electron relaxation appears or dissapears.

The dependence of the thermoelectric power and resistivity of the samples with different oxygen contents on a gydrostatic pressure up to 12 kbar was measured at room temperature. The dependence is found to be approximately linear. A slope of the resistance (thermoelectric power) vs. pressure dependence increases (decreases) as oxygen content diminishes from 7 to 6.2. Figure 4 exhibits the derivatives $\frac{\partial \ln(\alpha)}{\partial P}$ and $\frac{\partial \ln(\rho)}{\partial P}$ vs. oxygen content. Such behavour of the pressure dependence remains unexplained and, probably, is due to the symmetry transition in the 1-2-3 system.

References

1) C.Uher and A.B.Kaiser, -Phys.Rev., 1987, B36, 5680.
2) A.Jezowski, J.Mucha, A.J.Zaleski, M.Ciszek, et al, -Phys.Lett., 1988, 127A, 225.
3) M.C.Karamargin, C.A.Reynolds, F.P.Lipschultz, P.G.Klemens, - Phys.Rev., 1972, B6, 3624.
4) J.G.Bednorz and K.A.Müller, -Z.Phys., 1986, B64, 189.
5) L.C.Bourne, A.Zettl, T.W.Barbee III, L.Cohen, - Phys.Rev., 1987, B36, 3990.
6) T.Hidaka, T.Matsui, Y.Nakagawa, -Jpn.J.Appl.Phys., 1988, 27, L553.
7) M.Gurvitch, A.T.Fiory, L.S.Schneemeyer, R.T.Cava, et al, -Physica, 1988, C153-155, 1369.
   A.Jezowski, K.Rogacki, R.Horyn and J.Klamut, -Ibid.,p.1347.
8) K.J.Leary, H.C.Loye, S.W.Keller, T.A.Faltens, et al, -Phys.Rev.Lett., 1987, 59, 1236.
9) S.C.Lee, J.H.Lee, B.J.Suh, S.H.Moon, C.J.Lim and Z.J.Khim, -Phys.Rev., 1988, B37, 2285.
10) J.Heremans, D.T.Morelly, G.W.Smith, and S.C.Strite III, - Ibid., p.1604.
11) V.Bayot, F.Delannay, C.Dewitte, J.-P.Erauw, et al, -Sol.St.Comm., 1987, 63, 983.
12) F.Herman, R.V.Kasovski, and W.Y.Hsu, -Proc.Intern.Workshop on Novel Mechanism of Superconductivity (1987, Berkeley, Calif., USA),- Plenum Press, New York, 1987, p.521.
13) N.P.Ong, Z.Z.Wang and J.Clayhold, -Ibid., p.1063.
14) J.M.Tarascon, L.H.Greene, B.G.Bagley, W.R.McKinnon, et al, -Ibid., p.720.
15) E.S.Itskevich, A.N.Voronovskii, A.F.Gavrilov and V.A.Sukhaparov, -Cryogenics, 1967, 7, 359.
16) Y.Ishizawa, O.Fukunaga, H.Nozaki, T.Tanaka, A.Ono, - Physica, 1987, B148, 315.
17) K.Mendelsohn, J.Olsen, -Proc.Phys.Soc., 1950, 63A, 2.
18) B.T.Geylikman, V.Kresin, -Zh.Eksp.Teor.Fiz., 1959, 36, 959. (Sov.Phys.JETP, 1959, 36(9), 677).
19) A.Junod, -Physica, 1988, C153-155, 1078.
    K.S.Gavrichev, V.E.Gorbunov, et al, -Izv.Akad.Nauk., Neorg.Mater.(USSR), 1988, 36, 343.
20) V.I.Makarov, N.V.Zavaritsky, V.S.Klochko, et al , -Pis'ma v Zh.Eksp.Teor.Fiz., 1988, 48, 326.
    M.Suzuki, Y.Okuda, I.Iwasa, A.J.Ikushima, et al, - Jpn.Journ.Appl.Phys., 1988, 27, L308.
21) K.Nakamura, H.Aoki, A.Matsushita, et al, -Physica, 1987, B148, 318.
    B.Batlogg and R.J.Cava, -Ibid., p.173.
    P.Monod, M.Ribault, F.D'Yvoire, et al, -J.Physique, 1987, 48, 1369.
22) Z.Henkie, Z.Bukowski, R.Horyń, et al, - Sol.St.Comm., 1987, 64, 1285.
23) Z.Henkie, P.J.Markowski,R.Horyń, et al, - Phys.Stat.Sol., 1988, B146, 131.
24) V.I.Tsidilkovsky, I.M.Tsidilkovsky, -Fizika Metallov i Metalloved , 1988, 65, 83.
25) I.M.Lifshitz, -Zh.Eksp.Teor.Fiz., 1960, 38, 1569.   (Sov.Phys.JETP, 1960, 11, 1130).
26) V.G.Vaks, A.V.Trefilov, S.V.Fomichov, -Zh.Eksp.Teor.Fiz., 1981, 80, 1613.
    (Sov.Phys.JETP, 1981, 53(4), 830).

THE EFFECT OF A HIGH PRESSURE UP TO 210 KBAR UPON THE SUPERCONDUCTIVITY OF MONOCRYSTALLINE AND CERAMIC SPECIMENS OF $YBa_2Cu_3O_{7-y}$

Berman I.V., Brandt N.B., Romashkina I.L., Sidorov V.I.
(Moscow, Moscow State University, USSR)
Han Cui yng (Beijing, Institute of physics
Chinese Academy of Sciences)

1. Since the publication of the pioneering work of Chu et al. [1], the study of the behaviour of hightemperature superconductors under high pressures has been of great interest. Most of such studies, however, have been made in the range of relatively low (up to 20 Kbar) pressures (to 40 Kbar in ref. [2]) so that the question of how the critical temperature T_c behaves in the range of much higher pressures remains, to a certain extent, open. An attempt to study $La_{1.8}Sr_{0.2}CuO_{4-y}$ under pressure up to 160 Kbar [3] turned out unsuccessful, for the semiconductor behaviour of the electrical resistance R had been observed in the entire temperature range from 300 to 4.2 K. In the system $YBa_2Cu_3O_{7-y}$, great discrepancies have been observed not only in the value of the derivative dT_c/dP but also in its sign -indeed, $dT_c/dP > 0$ in [4], but < 0 in [5].

In the range of hydrostatic pressures up to 20 Kbar, one can observe fully completed transitions to the superconducting (SC) state, their width increasing somewhat under compression. The situation is changed when one goes over to the pressure range above 50 Kbar, where Bridgman anvils, imposing severe limitations on the size of the specimens, are used. The width of SC transitions is here greater than in the hydrostatic pressure range, the transitions being sometimes observed [3, 6, 7] against the background of an increasing (or a very slightly decreasing) resistance when the specimens are cooled.

The interpretation of the data obtained in studying $YBa_2Cu_3O_{7-y}$ ceramics by imploying Bridgman anvils [3, 6] is complicated by the fact that the measurements were made by the two-terminal technique, when the value of the measured resistance includes also the resistance of the anvils [6] or of the brass foil [3] used as one of the electrodes. The data of these works preclude therefore any definite conclusion about the completedness of SC transitions.

Completed transitions to the SC state under high pressure were observed in refs. [8, 9]. However, their data are significantly different: in [8], the T_c of $YBa_2Cu_3O_{7-y}$ increases from 95 to 107 K when the pressure is raised from 0 to 149 Kbar, whereas in [9] where single crystals of $YBa_2Cu_3O_{7-y}$ were used, it passes, as the pressure is raised in the range 0 to 100 Kbar, through a gently sloping peak at ∼ 50 Kbar.

In the present work, the dependence of R on T for single-crystal and ceramic specimens of $YBa_2Cu_3O_{7-y}$ has for the first time been investigated in the temperature range from 300 to 1.5 K and a wide pressure range up to 210 Kbar [7, 10].

2. Experimental procedure

Pressure was set up at room temperature in a high-pressure chamber (fig.1), compressed between the planes of Bridgman anvils made of polycrystalline diamonds [11]. The high-pressure chamber consists of two disks and two guard rings ∼ 20 mu thick, pressed out of finely powdered Fe_2O_3, and two steatite washers, with the diameter equal to the inner diameter of the guard rings. Specimens 30 by 30 by 300 mu in size were put between the steatite washers. The anvil diameter varied from 1.2 to 1.8 mm, and the diameter of the steatite washers, from 0.6 to 0.4 mm.

The electrical resistance of the specimens was measured by the four-terminal technique, with platinum strips ∼10 mu thick serving as the electrodes. To create pressure, a mechanical press or a smallsize multiplicator were used, which made it possible to make measurements on the same specimen as the pressure was raised or lowered. In this way, it was possible, on the one hand, to accurately determine the relative magnetude of a change in pressure and electrical resistivity and, on the other hand, to study the reversibility of the effects observed. The value of the pressure was found to within 5% from the shift

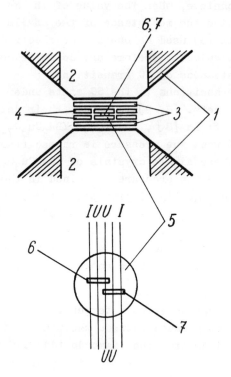

Fig.1. High pressure chamber: 1,2 - anvils, 3,4 - Fe_2O_3 disks and guard rings, 5 - steatite washers, 6 - samples, 7 - lead gauge, I, U - electrodes.

in the T_c of a lead pressure gauge placed in the chamber together with the specimen.

In the measurements, a freshly chipped - off specimen was put in the high-pressure chamber, which was immediately compressed up to 30 ÷ 40 Kbar. The chamber then becomes hermetically sealed, thus preventing the condensation of water vapour and liquid gases upon the specimen surface, as well as any change in the concentration of oxygen during evacuation and in the course of measurements.

3. Curves of superconducting transitions. The effect of specimen deformation resulting from nonhydrostatic pressure.

The curves of transitions to the SC state in ceramic specimens of $YBa_2Cu_3O_{7-y}$ under various pressures are presented in fig.2. For graphicness the origin on the temperature scale is displaced by 10 K for each curve, T_c values being indicated by the arrows. It can be seen that, although R drops to 0 in a transition to the SC state, the transitions themselves are greatly smeared. The region of a relatively rapid linear decrease of R is followed by a smeared transition segment of several tens of degrees. As the pressure increases, the slope of the linear part steepens (SP transitions broadening), while the smeared transition part somewhat narrows. At P < 100 Kbar, a SP transition is preceded by an increase in resistivity and at P > 100 Kbar, R slightly decreases upon cooling.

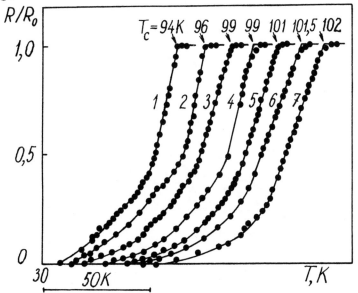

Fig.2. Superconducting transitions in ceramic samples of $YBa_2Cu_3O_{7-y}$ at various pressure, Kbar: 1-61, 2-67, 3-87, 4-97, 5-119, 6-140, 7-170.

Obviously, an interpretation of the data on the effect of a high pressure upon the superconductivity of ceramic high-temperature superconductors may be complicated by the presence in them of intergranular boundaries. More reliable results may be obtained with monocrystalline specimens.

Fig.3 shows the curves for SC transitions in single crystal $YBa_2Cu_3O_{7-y}$ at various pressures. In the range of relatively low pressures, the resistivity of single crystal specimens decreases very abruptly, byt not down to zero, in a SC transition. As the pressure is raised, the slope of the linear parts of SC transitions remains practically unchanged over a wide pressure range ($30 < P < 100$ Kbar), while SC transition become fully completed, shifting towards the higher temperature. The strongly smeared SC transition observed in $YBa_2Cu_3O_{7-y}$, both poly- and monocrystalline, appears to be the consequence

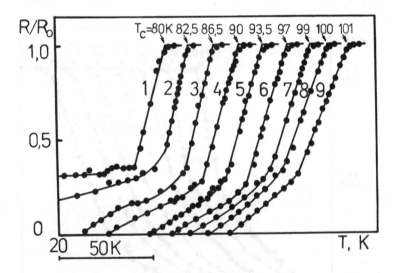

Fig.3. Superconducting transitions in single crystals of $YBa_2Cu_3O_{7-y}$ at various pressures, Kbar: 1-36, 2-46, 3-53, 4-75, 5-86, 6-93, 7-100, 8-120, 9-140

of a great nonuniformity (in T_c values) for small-size specimens, in which surface effects play the dominant role. The surface is known to be the most vulnerable part of superconductors, as all processes leading to structural disturbances (such as oxygen diffusion, incorporation of the vapours of water, carbon dioxide) initially occur on the surface. The high variations of the energy gap value across the surface of specimens that have been observed by the tunneling spectroscopy techinique in [12] as the probe moved across the surface seem to be due to exactly this circumstance.

In connection with this, there arise the question of which method of determining T_c in the given systems is the most correct.

The distinguishing feature of the distribution of regions according to the value of T_c in them (the distribution of volumes with different T_c) is that the above distribution has a very distinct boundary on the side of high temperatures, which is determined by T_c values for the specimen volumes with the most perfect structure, giving, in its turn, the critical temperature of a perfect material. In contrast, on the side of low temperatures, the distribution may have a long tail, which may extend down to very low temperatures, because a structural disturbance can lower T_c very significantly, even to a complete disappearance of the superconductivity [13].

It is therefore evident, then, that determining T_c from the value observed at the beginning of a SC transition (i.e., T_{co}) is physically meaningful, because this value is closest to that of perfect specimens, whereas the temperature found from the value observed at the end of a transition will correspond to reaching the percolation threshold (formation of an infinite cluster made up by SC regions), which emerges when a volume constituting about 20% of the specimen volume goes over to the SC state. Accordingly, the value of T_c found from the value observed in the middle, or at the end, of a transition is physically meaningless, because it should be considered rather as a characteristic distribution function for regions with different T_c, and not a characteristic of the superconducting properties of a material.

Under the influence of a pressure which is not hydrostatic (and this seems to determine the magnitude of shearing deformations) there may occur two processes: the number of high T_c regions decreases and regions with the lower T_c will appear. In this case, the procedure of determining T_c by its value observed at the beginning of a transition remains valid while there is a significant percentage of regions characterized by limitingly high T_c, although the width of a SC transition may increase due to the shift of the percolation threshold towards the lower temperatures. As the pressure becomes increasingly less hydrostatic, high - T_c regions begin to be destroyed and T_c begins to shift toward the lower temperatures. In this situation, the T_{co} parameter can no longer characterize the properties of a SC material.

And it is to the strong effect of nonhydrostatic deformations and the resulting destruction of SC regions that the vanishing of the superconductivity and the change in the character of the temperature dependences of the electrical resistivity of $La_{1.8}Sr_{0.2}CuO_{4-y}$ at P to 170 Kbar [3] appear to be due.

4. Dependences of T_c on pressure in poly - and monocrystalline $YBa_2Cu_3O_{7-y}$

Fig.4 illustrated the T_c as a function of pressure in ceramic $YBa_2Cu_3O_{7-y}$ specimens as the pressure is increased and then removed. The value of T_c corresponding to a transition in a massive sample is used as T_c at P = 0. A small irreversible change in T_c observed after the specimens are compressed to 200 Kbar ($\sim$ 2 K) is indicative of the absence of effects associated with a noticeable destruction of SC regions.

As the pressure is raised to 110 Kbar, the T_c of ceramic $YBa_2Cu_3O_{7-y}$ increases at a rate of $dT_c/dP \simeq 0.11$ K·Kbar^{-1}, which is greatly reduced as soon as P = 110 Kbar is reached. The nonlinear character of the dependence of T_c on P obtained in the present study differs from the linear dependences of T_c on P obtained for the same (ceramic) system in refs. [3, 6]

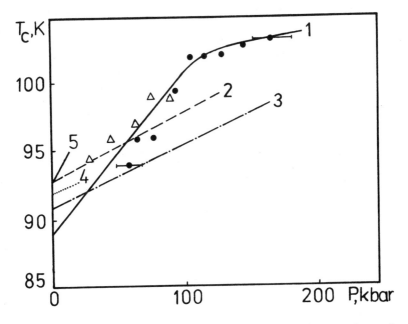

Fig.4. Pressure dependence of T_c in ceramic samples of $YBa_2Cu_3O_{7-y}$. Circles - increasing, triangles - removing pressure. 2 - data from Ref. [6], 3 - from Ref. [3], 4 - from Ref. [4], 5 - from Ref. [14].

(see curves 2 and 3 in fig.4). In the hydrostatic pressure range, the values of $dT_c/dP = 0.05$ K Kbar^{-1} are close to those obtained in [3, 6] (see curve 4 in fig.4). For the systems $(Y_xBa_{1-x})CuO_{2.3}$ at $X = 0.325$, 0.35 and 0.425, the authors of [14] obtained dT_c/dP values of 0.16, 0.1, and 0.1 K Kbar^{-1}, respectively. The discrepancies observed in dT_c/dP values of ceramic specimens of $YBa_2Cu_3O_{7-y}$ may be due to both composition variations and possible error in determining T_c arising from the deviation of compression conditions from hydrostatic (in all above-mentioned works SC transitions broaden under pressure). An exception is ref. [15], where a SC transition has been observed to narrow as a result of compression.

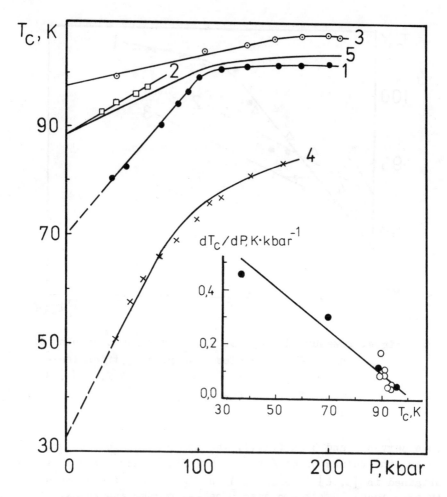

Fig.5. Pressure dependence of T_c in monocrystalline (curves 1-4) and ceramic (curve 5) samples of $YBa_2Cu_3O_{7-y}$.

Fig.5 shows T_c as a function of pressure in the $YBa_2Cu_3O_{7-y}$ system, constructed from the present data (curves 1 through 4 are for single crystals, 5 for ceramics). The data we have obtained enable us to suppose that the pressure dependence of T_c in a wide pressure range is as follows. As the pressure is raised, T_c linearly increases; then the rate of increase gradually reduces until, at some critical pressure (which is the lower the smaller the initial T_c at $P = 0$), it turns to zero. The question of how T_c behaves at high pressures remains, for the time being, open, although, in our opinion, the shape of the dependence under discussion (i.e. whether T_c retains a constant value or **whether**, passing through a peak, it decreases) is of the fundamental importance for the elucidation of the mechanism by which pressure affects the behaviour of T_c.

Another interesting feature observable in the P-dependences of T_c is that the value of the dT_c/dP for the orthorhombic phase of $YBa_2Cu_3O_{7-y}$ appears to be strongly dependent on the T_c value at $P = 0$, the specimens with the lower T_c having the largest dT_c/dP value. A tentative dependence of dT_c/dP on T_c for the various specimens is illustrated in the inset of fig.5.

Within the phonon mechanism of superconductivity, the change in T_c due to compression (a decrease, as a rule) in the nontransition metals and alloys can be explained by an increase in the phonon frequencies w (lattice stiffening). For the transition metals and alloys, the pressure induced change in T_c values may be due to the competition between two contributions: an increase in T_c resulting from an increase in the ionic potential and a decrease of T_c caused by an increase in phonon frequencies. One may then observe either an increase or a decrease of T_c under compression, while characteristic values of dT_c/dP lie within $\pm$ 0.01 K Kbar, except for La and the high-pressure phases of Ba and Y, for which dT_c/dP = 0.1, 0.05 and 0.03 K Kbar, respectively.

The measurements of phonon frequencies at pressures up to 16 Kbar have shown [16] that the phonon modes with the frequencies 145, 300 - 340, 440 and 506 cm^{-1} linearly increase in frequency under compression. This increase corresponds to the Gruniesen parameter $\gamma_G = -\frac{d \ln w}{d \ln V} = 1 \div 2$ (V being the volume) lying in the range of values characterizing typical transition metals. The compressibility values found in [17] are also close to those characteristic of the transition metals. It thus appears that the change in phonon frequencies and ionic potential can not account for the observed dT_c/dP values in $YBa_2Cu_3O_{7-y}$, which by nearly an order of magnitude exceed those in most transition metals; nor can it explain the correlations between T_c and dT_c/dP.

One possible reason for the latter may be the redistribution of oxygen between the various sites in the lattice, as a result of which the concentration of oxygen in layers responsible for superconductivity increases. Obviously, the redistribution should lead to a strong dependence of T_c upon P in specimens originally deficient in oxygen, while in those with the originally optimal oxygen concentration there should be no such dependence. Unfortunately, structural data which would testify to the existence of the redistribution effect have not yet been available. It is possible, that the weakening of localization effects under pressure plays an important role in the nature of T_c increasing in single crystals $YBa_2Cu_3O_{7-y}$ with low T_{co} [18].

The aforesaid suggests that it should be of special interest now to study the pressure dependence of T_c in the range of the still higher pressure and also in single crystals with originally low (due to oxygen deficiency) T_c values.

REFERENCES

1. Chu C.W., Hor P.H., Meng R.L., Gao L. and Huang Z.J. - Science, 1987, v.235, pp.567 - 569
2. S.Yomo, C.Murayama, T.Takanashi, N.Mori, K.Kishiom, K.Kitazawa and K.Fueki. - Jap.J.Appl.Phys., 1987, v.26, L603
3. A.Driessen, R.Griessen, H.Hemmes, N.Kocman, J.Ractor and P.H.Kes. - Phys.Rev.B, 1987, v.36, p.5602
4. Yomo S., Murayama C., Utsomi W., H.Takahashi, T.Yagi, N.Mori, T.Tamegai, A.Watanabe and Y. Iye. - Jap.J.Appl. Phys., 1987, v.26, supplement 26 - 3, pp.1107 - 1108
5. Murata K., Ihara H., Tokumoto M., M.Hirabayashi, N.Terada and K.Senzaki. - Jap.J.Appl.Phys., 1987, v.26, N 4, pp. L471 - L472
6. D.Odai, K.Takahashi and M.Ohta. - Jap.J.Appl.Phys., 1987, v.26, L820
7. Berman I.V., Brandt N.B., Kystaubaev T.Z., Kozlov R.I., Sidorov V.I., Han Cui yng, Graboi I.E. and Kaul A.R. - In: "Problems of High-Temperature Superconductivity", Sverdlovsk, 1987, Part II, p.213
8. Maple M.B., Dalichaouch Y., E.A.Early; J.M.Ferreira, R.R. Hake, B.W.Lee, .T.Markert, M.W.Moeifresh.
   - Proceedings of ICHTS, 1988, Switzerland, B232
9. Latter N., Wittig J. - Proceedings of ICHTS, 1988, Switzerland, C218
10. Berman I.V., Brandt N.B., Graboi I.E., Kaul A.R., Kozlov R.I., Romashkina I.L., Sidorov V.I. and Han Cui yng. - Pis'ma v ZhETF, 1988, v.47, ess.12, pp.634-637
11. Brandt N.B., Berman I.V., Kurkin Yu.P. and Sidorov V.I. - Cryogenics, 1976, v.50, p.47 - 49
12. Volodin A.P., Khaikin M.S. - Pis'ma v ZhETF, 1987, v.46, iss.11, pp.466 - 468
13. Kobelev L.Ya., Nugaeva L.L., Babushkin A.N., Lobanov Yu.A., Yakovlev E.N. and Babushkina G.V. - In: "Problems of High-Temperature Superconductivity", Sverdlovsk, 1987, Part II, p.215

14. Yoshida H., Morita K., Noto K., Kaneko T. and Fujumori H.
    - Jap.J.Appl.Phys.,,1987, v.26, L1867
15. Bud'ko S.L., Gapotohenko A.G., Golovashkin A.I., Ivanenko
    O.M., Itskevich E.S. and Mitsek K.V.

    - Pis'ma v ZhETF, 1987, v.46, iss.9, Suppl. pp.276 - 277
16. Syassen K., Hanfland M., Strossner K., Hotz M., Kiess W.
    and Gardona M. - Proceedings of ICHTS, Switzerland, 1988,
    A188
17. Aleksandrov I.V., Goncharov A.F., Stishov S.M. - Pis'ma
    v ZhETF, 1988, v.47, iss.7, pp.357 - 360
18. Berman I.V., Brandt N.B., Kostyleva I.E., Pavlov S.K.,
    Sidorov V.I., Chudinov S.M. - Pis'ma v ZhETF, 1986, v.43,
    iss.1, pp.48 - 50

ACOUSTIC CHARACTERISTICS AND PECULIARITIES OF THE LATTICE VIBRATION SPECTRUM IN $La_{2-x}Sr_xCuO_4$ (x=0;0,2) AND $YBa_2Cu_3O_y$ (y=6;7)

V.S.Klochko*, V.I.Makarov*, V.F.Tkachenko**, A.P.Voronov**, N.V.Zavaritsky***

* Institute of Physics and Technology, the Ukrainian SSR, Academy of Sciences, 310108 Kharkov, USSR
** Institute of Single Crystals, 310072 Kharkov, USSR
*** Institute for Physics Problems, Academy of Sciences of the USSR, 117334 Moscow

Temperature variation of the longitudinal sound absorption, $\alpha(T)$, and its velocity, $S(T)$, in the compounds of $La_{2-x}Sr_xCuO_4$ and $YBa_2Cu_3O_y$ is of the same kind for (x=0;0,2) and (y=7;6). Anomalies in the temperature dependence of α and S are associated with the sound soft optical mode interaction.

At present, there is no common viewpoint on the temperature dependence peculiarities of acoustic characteristics in the high-temperature superconductors. This is primarily due to contradictions of experimental data[1-10]. For example, although the significant anomalies in the sound velocity S or absorption α were observed near T_c in some studies[5,6], they were not found in the others. An ambiguous interpretation of the results obtained was explained also by the fact that the measurements were usually carried out on isolated samples thereby it made choice among different models difficult. Below the results of acoustic characteristic study in $YBa_2Cu_3O_y$ at y=7;6 and $La_{2-x}Sr_xCuO_4$ at x=0,2;0 ceramic samples and a single crystal sample La_2CuO_4 as well are given. Similar complex study has not been carried out before.

The measurements were performed at the frequencies of 20 and 50 MHz using the standard pulse technique with a pulse length 1-1,5 μsec. Electronic signal-comparison circuit allowed the measurement of the sound velocity with a relative accuracy $\sim 10^{-2}$ per cent and the absorption coefficient with that of $\sim 10^{-1}$ db/cm. At the temperatures above 150 K, the measurement accuracy decreased as the temperature behaviour of the delay line had to be taken into account. Samples available in the form of disks 1,5-3 mm thick were protected against contract lubricant penetration into the bulk of material. During thermocycling from 4,2 to 250 K, neither hysteresis effects nor discontinuities in the sound velocity and absorption were observed. A study was made of textured ceramic samples with the c-axis directed along the sound propagation. In the case of $YBa_2Cu_3O_y$ a single sample was used for which the magnitude of y was varied by vacuum annealing. A plate of 3x3mm size, 2.5mm thick was cut normal to the c-axis from the starting single crystal La_2CuO_4 of orthorhombic structure with axes a = 5.3845 Å, b =5.3528 Å and c=13.1490 Å, the disorientation of blocks being less than 5'.

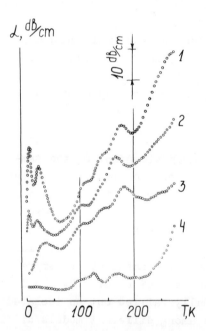

Fig.1. Temperature dependence of sound absorption: La_2CuO_4 (ceramics) 1-50MHz, 2-20MHz; 3-$La_{1,8}Sr_{0,2}CuO_4$, 50 MHz; 4-La_2CuO_4 (single crystal), 50 MHz, sound directed along the c-axis.

In the samples of lanthanum system one can observe peculiarities in

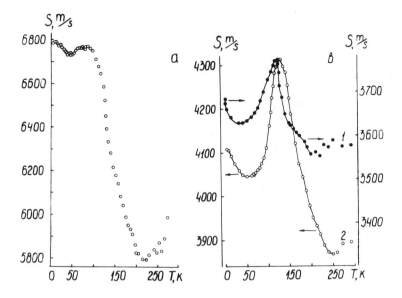

Fig.2. Sound velocities in the lanthanum system:
a-single crystal La_2CuO_4, 50 MHz; b-$La_{2-x}CuO_4$
ceramics, 1 (dots) - La_2CuO_4
2 (circles) - $La_{1,8}Sr_{0,2}CuO_4$

orthorhombic modification and at y=6 it has a dielectric tetragonal one. In this experiment, one and the same sample was and hence, the complicating factors had little effect. The orthorhombic to tetragonal structural phase transition is attended with an increase in the sound velocity by 10-12 per cent and reduction of $\alpha(T)$ and $S(T)$ the temperature varies in the range of 4,2 to 150 K (Fig.3). Similar results were obtained in[10] but it has become known to us only while writing this article. Note, that here again the only quantitative variation of the $\alpha(T)$ and $S(T)$ dependences is obtained.

The whole complex of these results show that the peculiarities of the acoustic characteristics are not effected the curves of the sound absorption temperature dependence, $\alpha(T)$ at $T \gtrsim 50$ K, which are absorption peaks at the temperatures 100-120 K and 160-170 K (Fig.1), their magnitude being 5-10 per cent of the temperature-independent background (~ 50 db/cm). Decrease of the sound frequency slightly shifts the peak to low-temperatures (compare curves 1 and 2). Besides, at $T < 25$ K the absorption coefficient is approximately $\sim \omega$ rather than $\sim \omega^2$ as is the case for similar materials.[12,13] The stronthium-bearing sample is a superconductor with $T_c = 38$ K. At $T \approx T_c$ it does not reveal any absorption peculiarities (curve 3), in particular, there is no decrease in α at $T \lesssim T_c$ which is typical for ordinary superconductors.

At the temperatures $T \lesssim 50$ K, the $\alpha(T)$ dependence of the La_2CuO_4 ceramic sample is significantly different from that of a single crystal (Fig.1). This difference seems to be caused by a crystallographic anisotropy of the absorption. For example, in the range of 6-30 K the anomalies similar to the ones obtained in the ceramic sample were also observed in a large-block single crystal when the sound was propagating at an angle to the c-axis.

As the temperature decreases below $T \sim 200$ K the sound velocity shows a marked increase by ~ 10 per cent and reaches its peak at $T \sim 120$ K in lanthanum samples (Fig.2), then in the vicinity of 50 K there is a drop in the sound velocity by 1 per cent in a single crystal and by 5-6 per cent in ceramic samples.

This variation of $S(T)$ is of the same kind both in superconducting and normal ceramics, therefore, it is unreasonable to consider the one as a "precursor" of the superconducting transition. Similar characters of $\alpha(T)$ and $S(T)$ variations in dielectric and metal-like samples provides some evidence for a small contribution of the conduction electrons into acoustic characteristics of lanthanum compounds. The value of the sound attenuation due to electrons previously reported by us[3] was probably overestimated.

At $y = 7$, $YBa_2Cu_3O_y$ compound has a superconducting

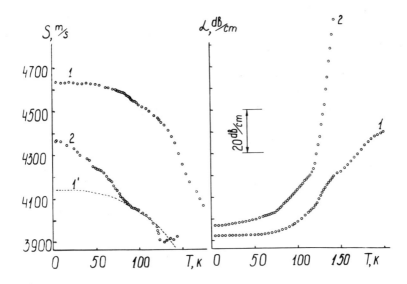

Fig.3. Acoustic characteristics in $YBa_2Cu_3O_y$, ceramics 1 and 1'-y=6; 2-y=7. The curve 1' displays a relative variation in S.

by a variation of the composition of the studied compounds. This favours the assumption that the potential structural transitions do not play a leading role in the occurence of the peculiarities in the $\alpha(T)$ and $S(T)$ dependences. However, as it follows from the neutron measurements[11] the composition of the studied compounds has but a little effect on the lattice vibration spectrum. The reason for the anomalies in $\alpha(T)$ and $S(T)$ should be probably sought in the character of the sound-intrinsic lattice vibrations interaction in the studied compounds. As it is known[12] the nonequilibrium process cause the sound absorption due to a redistribution of the thermal energy between different modes of vibrations if the vibration frequency changes due to the lattice sound deformation. The magnitude of absorption is determined by multiphonon processes of energy exchange between different vibration modes which are energy, pulse and vibration polarization resolved. In compounds with a perovskite lattice[13], soft optical modes whose energy is changed by the deformation seem to be responsible for the sound absorption.

In a "lanthanum" system this is primarily an optical mode with $\mathcal{E}_0$ =10 MeV, its existence being supported by neutron experiments[14]. At higher energies, these are the modes with $\mathcal{E}$ =30 and 45 MeV, respectively. The presence of these modes is inferred both from the calculation of dispersion curves for the compound[15] and optical data[16]. A shift of the characteristic energies in the spectrum with a temperature variation obtained in this work is an additional evidence of the fact that these modes are "soft". Note that all these modes involve the vibrations of oxygen complexes in different combinations.

The sound-soft optical mode interaction also causes a variation in the sound velocity. Gradual "freezing out" of the optical vibrations with a reduction in temperature leads to the increase in the sound velocity as a result of enhanced stiffness of the lattice. From this viewpoint, a variation of dependences $\alpha(T)$ and $S(T)$ in "yttrium" ceramics can be naturally associated with a respective decrease of the spectral density of vibration modes in the oxygenized complexes which are responsible, in particular, for the temperature dependences of these quantities.

The whole complex of the results obtained can be thus explained, at least qualitatively, by the sound-soft optical mode interaction in the compound studied. The quantitative description is complicated by the presence of several soft modes, as well as the necessity of taking into account the anisotropy in the characteristics of compounds described in the paper.

The authors wish to express their thanks to Dr. V.P.Seminozhenko for his support.

References

1. K.Fossheim, T.Laegreid, E.Sandvold, F.Vassenden,K.A.Muller, I.G.Bednorz. Sol.Stat.Com. 63, 531 (1987).
2. A.Migliori, Ting Chen, B.Alavi and G.Gruner. Sol.Stat.Com. 63, 827 (1987).
3. V.I.Makarov, V.S.Klochko, N.V.Zavaritsky, S.V.Petrov. Pis'ma Zh.Eksp.Teor.Fiz. 46, 156 (1987).
4. V.G.Bar'yakhtar, V.N.Pan, V.F.Taborov, V.F.Tarasov, N.V.Shevchyk, O.I.Shulenkova, A.I.Tcherbak. FNT 13,818 (1987).
5. A.I.Golovashkin, V.A.Danilov, O.M.Ivanenko, X.V.Mitchen, I.I.Perepechko, Pis'ma Zh.Eksp.Teor.Fiz.46, 273 (1987)
6. Wang Yening, Shen Huimin, Lhy Jinsong et al. J.Phys.C.Sol.Stat.Ph.20, 665 (1987).
7. D.J.Bishop, P.L.Gammel, A.P.Ramires, R.J.Cava. Phys.Rev.B 35, 8788 (1987).
8. D.J.Bishop, A.P.Ramires, P.L.Gammel et al. Phys.Rev.B 35, 2408 (1987).
9. M.-F.Xu, H.-P.Baum, A.Schenstrom, Bimal X.Sarma and Moises Levy. Phys.Rev.B. 37, 3675 (1988).
10. Mazaru Suzuki, Yuidi Okuda et al. Jap.J.of Appl.Phys. 27, 308 (1988).
11. B.Renker, F.Gompf, E.G ring et al. Z.Phys.B.Condens.Matter 65, 15(1987).
12. E.I.Liphshits, L.P.Pitaevsky, Fizicheskaya Kinetika,Moscow, "Nauka", 72, 1979.
13. G.Barret. "Fizicheskaya Acoustika" edited by Y.Mezon, Mir Publisher, Moscow, v.YI, p.90, 1973.
14. R.J.Birgenau, C.J.Chen, D.R.Gable, H.P.Iensen, M.A.Kastnee, C.I.Peters, P.I.Picone, Tineke Thio, T.R.Thurston and H.L.Tuller. Phys.Rev.Lett. 59, 1329 (1987).
15. Y.Horie, F.Fukemi, S.Mase.Jap.J.Appl.Phys.26, 2623 (1987).
16. S.Blumenroeder, E.Zirngeebl, J.D.Thompson, P.Killough, J.L.Smith and Z.Fisk. Phys.Rev.B 35, 8788 (1987).

EFFECTS OF LOCALISATION IN ATOMIC-DISORDERED HIGH T_c SUPERCONDUCTORS

B.N.Goshchitskii, S.A.Davydov, A.E.Karkin, A.V.Mirmelstein, M.V.Sadovskii, V.I.Voronin

Institute for Metal Physics and Institute of Electrophysics, USSR Academy of Sciences, Ural Branch, 620219 Sverdlovsk GSP-170, USSR

ABSTRACT

The influence of disordering induced by fast neutron irradiation at 80 K on physical properties of the oxidic ceramics $RBa_2Cu_3O_{7-\delta}$ (R=Y,Ho,Er), $La_{1.83}Sr_{0.17}CuO_{4-y}$, Bi-Sr-Ca-Cu-O and a single crystal $YBa_2Cu_3O_{7-\delta}$ has been investigated. The obtained results show that localisation states exist in the system even at small disorder. The effects are interpreted in terms of the localisation theory.

1. INTRODUCTION

A comprehensive study of new high-temperature superconductors (HTSC) started after their discovery in 1986 [1] allowed to establish important experimental facts characterising superconductivity of these oxides. Fast neutron irradiation is one of the most developed methods used to affect ordered compounds in order to obtain unique information about their properties. This method allows to carefully introduce defects macroscopically homogeneously distributed over the volume of the sample to study the behavior of T_c and other properties during disorder with high reproducibility of results, the stoichiometry being unchanged. As known decrease in the electron free path ℓ with increasing concentration of defects does not itself affect the value of T_c in ordinary wide-band superconductors [2] . However, in A-15 compounds an essential variation of T_c during disorder may be explained by variation of such parameters as $N(E_F)$ (the density of electron states at Fermi level) and $\langle\omega\rangle$ (typical phonon frequencies) due to substantial rebuilding of a crystal structure at high defect concentration. Electrical resistivity in this case shows the behavior characteristic of metals, i.e.: for small fluences of fast neutrons ϕ residual electrical resistivity ρ (o) grows proportionally to ϕ , while for high ϕ it saturates at the values of the order of 150-300 ohm·cm. The derivative $d\rho/dT$ decreases with increasing $\rho(o)$ and when ℓ is close to interatomic distance $\rho(T)$ practically does not depend on temperature [3] . This is not the case for HTSC. Disorder caused by neutron irradiation decreases T_c and leads to qualitatively different changes in electrical resistivity, viz.: linear dependence of electrical resistivity ρ (T) transforms into exponential characteristic of "jump-like" conductivity by localised states [4] . The first investigations on HTSC irradiated by fast neutrons have already shown [4-7] that disorder alters their properties in such a combination that has never been observed experimentally.

Here in terms of the localisation theory we'll discuss the previous [4-7] and new results including BiSrCaCuO (zero resistivity at T=70 K, $\Delta T_c \simeq 10$ K, $\rho_{120}=0.95 \times 10^{-3}$ ohm·cm, $\rho_{300}/\rho_{100} = 1.8$), $RBa_2Cu_3O_{7-\delta}$, where R=Ho,Er, and a single crystal $YBa_2Cu_3O_{7-\delta}$ ($T_c \simeq 80$ K, $\rho_{300} \simeq 2$ mohm·cm) as well as effects of low-temperature annealing on T_c and electrical resistivity of HTSC disordered under fast neutron irradiation ($T_{irr} = 80$ K).

2. EXPERIMENTAL RESULTS

According to X-ray and neutron diffraction analysis on introduction of radiation defects $YBa_2Cu_3O_{6.95}$ shows partial oxygen redistribution over O4 and O5 sites, with the Debye-Waller factor and lattice parameters being increased [6]. Superconductivity is fully suppressed in the samples with essential orthorhombicity, (b-a)=0.05 Å. Thus, the ortho-tetra transformation itself is not responsible for decrease in T_c. It is likely that essential atomic displacements (static and dynamic) from their regular lattice sites result in appearance of a chaotic potential that is responsible for alteration of physical properties of HTSC during disorder. T_c and $\rho(T)$ change in the same way during disorder in all HTSC under study. Introduction of defects leads to a rapid decrease in T_c and broadening of superconducting (SC) transitions (Figs.1,2).

In the irradiated samples T_c partially reestablishes after annealing at 200 K and 300 K for 20 min. Week annealing at

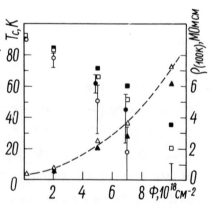

FIGURE 1. Dependence of T_c determined by resistivity measurements (T_c^ρ) and ac-susceptibility (T_c^χ) (the left scale) and electrical resistivity at T = 100 K, ρ(100 K) (the right scale) on fluence of fast neutrons for $YBa_2Cu_3O_{6.95}$. T_c^χ after irradiation at 80 K (o) and after annealing at 300 K for two weeks (●); vertical lines show the transition width; T_c after annealing at 300 K for two hrs (□) and two weeks (■); ρ(100) after annealing at 300K for two hrs (△) and two weeks (▲).

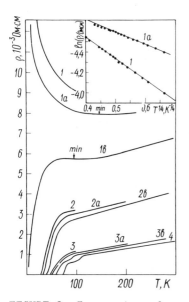
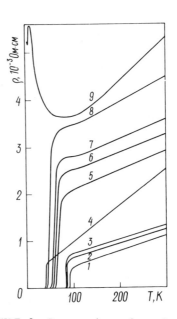

FIGURE 2. Temperature dependence of ρ for BiSrCaCuO irradiated by fast neutrons of fluence $7 \times 10^{18} cm^{-2}$ (curve 1), $3 \times 10^{18} cm^{-2}$ (curve 2), $2 \times 10^{18} cm^{-2}$ (curve 3) and $\phi = 0$ (curve 4). Indexes a and b correspond to annealing at 200 and 300 K for 20 min, respectively.

FIGURE 3. Temperature dependence of ρ for YBaCuO (curves 1-3, 5-8) and LaSrCuO (curves 4,9) irradiated by fast neutrons. 1) $\phi = 0$. 3)6)8) irradiation at 80 K by fluences 2, 5 and $7 \times 10^{18} cm^{-2}$, respectively, plus annealing at 300K for two hrs. 2)5)7) irradiation by a fluence of 2, 5 and $7 \times 10^{18} cm^{-2}$, respectively, plus annealing at 300 K for two weeks. 4) $\phi = 0$. 9) irradiation by a fluence of $5 \times 10^{18} cm^{-2}$ plus annealing at 300 K for two hrs.

room temperature results in further recovery of T_c and narrowing of SC transition (Figs.1,2). The superconductors $RBa_2Cu_3O_{6.95}$ (R=Er,Ho) and Bi-Sr-Ca-Cu-O irradiated by a fluence of $7 \times 10^{18} cm^{-2}$ do not show superconductivity up to 1.7 K while $\rho(T)$ vs. temperature curve points to localisation of carriers. SC transition appears after annealing (Fig.2 for SiSrCaCuO) at

300K while non metallic dependence $\rho(T)$ ($d\rho/dT < 0$) still remains. In $La_{1.83}Sr_{0.17}CuO_{4-y}$ and $YBa_2Cu_3O_{6.95}$ the value of the derivative of the upper critical field H'_{c2} remains the same (within the experimental uncertainty connected with broadening of the transitions) with increasing by order. Note that in metallic ordered compounds increasing electrical resistivity should lead to increase in H'_{c2} observed previously in the irradiated A-15 [3] and Shevrel phases [8].

At small fluences of fast neutrons $\phi < 5 \times 10^{18} cm^{-2}$ $\rho(T)$ remains linear for $T > T_c$. For $(5-10) \times 10^{18} cm^{-2}$ the $\rho(T)$ curves are characterised by simultaneous presence of the fractions with $d\rho/dT > 0$ and $d\rho/dT < 0$ at high and low temperatures, respectively. For $\phi > 1 \times 10^{19} cm^{-2}$ $\rho(T)$ is described by the known formula

$$\rho(T) = A \exp(Q/T^{1/4}), \quad Q = 2.1 [N(E_F) R_{loc}^3]^{-1/4}. \quad (1)$$

Radiation effects in HTSC induced by fast neutron irradiation at t=80K are relatively unstable, i.e.: isochronous annealing for 20 min at T=200 and 300K leads not only to partial recovery of T_c but also of $\rho(T)$ (Figs.2,3). Fig.4 depicts exponential growth of ρ_{80K} with increasing fluence (measurements

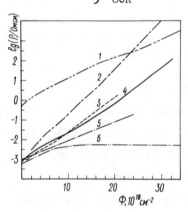

FIGURE 4. Dependence of $lg\rho$ on fluence obtained during irradiation at 80K. 1)La_2CuO_4; 2)$YBa_2Cu_3O_{6.95}$; 3)single crystal YBaCuO, ρ is measured perpendicular to the c-axis; 4)$La_{1.83}Sr_{0.17}CuO_4$; 5)BiSrCaCuO; 6) $SnMo_6S_8$.

are performed directly during irradiation). Here we present also the analogous dependence for $SnMo_6S_8$ electrical resistivity of which is proportional to fluence for small and saturates for large ϕ, respectively. The observed variation of ρ depending both on fluence and temperature may be described by the empirical formula [4]:

$$\rho(T) = f(T) \exp(a \, \phi/T^{1/4}). \qquad (2)$$

3. DISCUSSION

Though the dependence of type (1) is observed directly in the experiment only at large fluences (when T_c is either low or absent) localised states are likely to occur well before. Exponential growth of ρ (T) during irradiation beginning from the smallest fluences as well as invariability of the derivative H'_{c2}, inspite of fast growth of electrical resistivity, and, probably, Curie-Weiss contribution to magnetic susceptibility [6] may be naturally considered as the evidence for existence of the localised states even in a slightly disordered sample (since the nature of temperature linear dependence of $\rho(T)$ in the initial sample is not clear it can't be evidence for absence of the localised states). Hence, suppression of superconductivity may be assumed to be connected not with the appearance of the localised states itself but with appreciable decrease in localisation radius with increasing disorder in the system. In the system with strong two-dimensional anisotropy of conductivity the minimum value of metallic conductivity may be appreciably higher than that typical of the three-dimensional isotropic system, viz. $(3-5) \times 10^2$ $ohm^{-1}cm^{-1}$, and reach the value of $\geqslant 10^3$ $ohm^{-1}cm^{-1}$

for the compounds under study [9]. In terms of these estimates even the initial HTSC samples used in our investigations may be considered to close to Anderson transition.

In the isotropic three-dimensional system we have [10]:

$$-\frac{\sigma}{N(E_F)}\left(\frac{dH_{c2}}{dT}\right)_{T_c} = \begin{cases} \frac{8e^2}{\pi^2\hbar}\phi_0, & \sigma > \sigma^* \\ \frac{\phi_0}{2\pi}\frac{\sigma}{[N(E_F)T_c]^{1/3}} \sim \frac{\phi_0 \sigma}{2\pi T_c(\jmath_0 \ell^2)^{2/3}}, & \sigma < \sigma^* \end{cases} \quad (3)$$

Here H_{c2} is the upper critical field, ϕ_0 is the magnetic flux quantum, σ is conductivity, ℓ is the electron free path, $\sigma^* = \sigma_c (T_c/E_F)^{1/3}$ is the typical conductivity, $\sigma_c = e^2 p_F/\pi^3 \hbar^2$ is the minimal metallic conductivity by Mott [11]. Thus, in the vicinity of the metal-insulator transition, when $\sigma < \sigma^*$, the relation of Gor'kov (the upper expression in eq.(3)) is not valid, and further H'_{c2} does not depend on σ. Such a behavior is observed experimentally for ceramic samples [6]. As for the quasi-two-dimensional systems $(H_{c2}^{\|,\perp})'$ shows qualitatively the same behavior for $\sigma < \sigma^*$. Experiments on single crystals undertaken now will help us to elucidate this problem in details.

Anderson showed [2] that for a given pairing interaction near the Fermi level assuming that superconducting order parameter is self-averaging T_c does not practically depend on whether the states are delocalised or localised. The only restriction existing for localised states is connected with the known discreteness of the electron spectrum in the localisation region (repulsion level effect) [11]. The states close enough by energy are located on large distances from each other. Cooper pairing may take place only between the electrons with centres of localisation lying within the sphere of the order of localisation length R_{loc}. However, these states are splitted by energy to the value of order $[N(E_F)R_{loc}^3]^{-1}$ [11]. Apparently the condition should be held that the value of the superconducting gap Δ (at T=0) is essentially higher than the value of this splitting:

$$\Delta \sim T_c \gg [N(E_F)R_{loc}^3]^{-1}, \qquad (4)$$

i.e. the energy interval $\sim \Delta$ should contain a lot of discrete levels with the localisation centres being within the region R_{loc}. Eq.(4) is equivalent to the requirement that R_{loc} should be essentially higher than the typical size of Cooper pair in highly disordered system [10]. This is the qualitative criterion for existence of superconductivity in the Anderson insulator.

Using the experimental data on electrical resistivity of the radiation-disordered $YBa_2Cu_3O_{7-\delta}$ samples (for $\phi > 5 \times 10^{18} cm^{-2}$) and the empirical formula (2) for less fluences

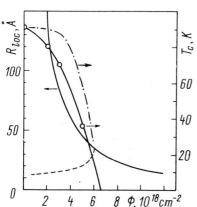

FIGURE 5. Dependence of T_c on fluence for YBaCuO (—o—). The solid line shows R_{loc}, calculated by formulas (1) and (2); dash-dot line shows T_c calculated by eq.(5a); dots show the minimal R_{loc} at which there may exist superconductivity with the given T_c according to eq.(3) (refer to the text).

one may calculate variation of R_{loc} depending on fluence. On the other hand, eq.(4) yields the limiting values of R_{loc} at which superconductivity may still exist in the system of localised electrons. Suppose that the left-hand side of eq.(4) equals $5T_c$ and for all fluences $N(E_F) = 5 \times 10^{33} (erg \cdot cm^3)^{-1}$ (here only the order of magnitude is of importance) one obtains the result shown in Fig.5. It is seen that the criterion of eq.(4) is not held for $\phi > (5-7) \times 10^{18} cm^{-2}$, being in good agreement with the experiment.

From the given estimates it is not clear why T_c should drop for high (in comparison with the limiting ones) values of R_{loc}. It is difficult to make a conclusion about the reason of T_c decreasing with increasing disorder because of lack of theoretical understanding of the nature of T_c in HTSC. Besides, in terms of the traditional considerations about the pairing interaction one may assume that suppression of T_c is connected with increasing effects of Coulomb repulsion in a single quantum (localised) state. According to [10] in the localisation region we have:

$$\ln \frac{T_{co}}{T_c} = \psi\left(\frac{1}{2} + \frac{\mu A_{E_F}}{4 T_c N(E_F)}\right) - \psi\left(\frac{1}{2}\right), \quad (5)$$

where ψ is the digamma function, μ is the Coulomb potential, $A_{E_F} \approx R_{loc}^{-3}$. For T_c calculations it is convenient to express the argument of ψ -function through resistivity using eq.(1):

$$\ln \frac{T_{co}}{T_c} = \psi\left(\frac{1}{2} + \frac{\mu T \left(\ln \frac{\rho(T)}{A}\right)^4}{4 T_c (2.1)^4}\right) - \psi\left(\frac{1}{2}\right). \quad (5a)$$

Now dependence of T_c on disorder may be easily calculated using the values of resistivity ρ (T=80K) from Fig.4 and assuming that $\mu \approx 1$ (Fig.5). The functional dependence $T_c(\phi)$ differs slightly from the experimental one (in the theory $dT_c'/d\phi \to 0$ for $\phi \to 0$). However, the qualitative agreement is out of question.

4. CONCLUSION

In terms of the localisation theory one manages to elucidate some typical features of the behavior of disordered HTSC. A number of the experimental facts, however, remains unexplained, for example, magnetic moments (although for small R_{loc} it is principally possible from the considered point of view), variation of spin-lattice relaxation rate on ^{63}Cu, heat capacity during disorder etc. [4,9]. Nevertheless, one may state that the fact that these materials are close to Anderson metal-insulator transition and existence of superconductivity with high enough T_c in the system of localised electrons is likely to be a specific feature of HTSC. This "non-metallic" behavior of these compounds becomes more evident when comparing, for example, with A-15 superconductors. With all their peculiarities the latter behave more like metals than HTSC. As for superconductors with the structure of the Shevrel phase, they present an intermediate case between the A-15 and HTSC, though they have lower T_c values than A-15.

REFERENCES

1. Bednorz J.G. and Müller. Z.Phys.B. 64 (1986) 189.
2. Anderson P.W. J.Phys.Chem.Solids 11(1959) 26.
3. Goshchitskii B.N., Arkhipov V.E., Chukalkin Yu.G.,Sov.Sci.Rev., Sec.A.: Phys.Rev. ed. by I.M.Khalatnikov, 8(1987) 520.
4. Davydov S.A., Karkin A.E. et al.Pis'ma JETP 47 (1988) 193.
5. Voronin V.I., Goshchitskii B.N., Davydov S.A. et al. Novel Superconductivity ed. by S.A.Wolf, V.Z.Kresin. N.Y.-London, Plenum Press (1987) 875.
6. Aleksashin B.A., Berger I.F. et al. Physica C. 153-155(1988) 339.
7. Voronin V.I., Davydov S.A. et al.Pis'ma JETP;Supplement 46 (1987) 165.
8. Davydov S.A., Arkhipov V.E., Voronin V.I., Goshchitskii B.N. Fizika metallov i Metalloved. 85(1983) 1092.
9. Lee P.A., Ramakrishnan T.V. Rev.Mod.Phys. 57(1985) 287. Abrahams E., Anderson P.W., Licciardello D.C., Ramakrishnan T.V. Phys.Rev.Lett. 42(1979) 673.
10. Bulaevskii L.N., Sadovskii M.V. Pis'ma JETP 39(1984) 524; J.Low Temp.Phys. 59(1985) 89.
11. Mott N.F., Davis E.A. Electron Processes in Non-Crystalline Materials (Clarendon Press, Oxford, 1979).

HIGH - T_c CERAMIC WEAK LINKS, rf-SQUIDs and THEIR APPLICATIONS

N.V.Zavaritsky and V.N.Zavaritsky
Institute for Physical Problems, USSR Academy of Sciences
117334 Moscow
General Physics Institute, USSR Academy of Sciences, 117942 Moscow

Abstract

123 : YBCO - ceramic "weak link" with dimensions of $\sim (10^{-2} cm)^3$ have displayed point contact Josephson junction characteristics in interference contour and under microwave radiation. Possibility of active devices realization with operation range up to $3.5 \cdot 10^{12}$ Hz at 77 K was confirmed by superior order Shapiro steps up to $n \gtrsim 100$ at 35.5 GHz observation on "weak links" characteristics. Rf-SQUIDs operated up to 90 K with external field sensitivity reduced by ambient noise to $2 \cdot 10^{-8} Oe \cdot s^{\frac{1}{2}}$ and flux noise of $5 \cdot 10^{-4} \Phi_0 \cdot s^{\frac{1}{2}}$ for single- and double-hole sensors at 77K correspondingly were produced and studied. The applications of these SQUIDs were briefly discussed and illustrated by ceramic shields and magnetic moment investigations.

1. Introduction

Recent discovery of ceramic oxide superconductors with critical temperatures well above that of liquid nitrogen has opened up the potential for use of superconductivity at readily accessible cryogenic temperatures. Possible applications of such a materials fall broadly into categories of power engineering and active devices. Realisation of devices based on a weakly coupled superconductors seems to be more attainable, for they do not need high values of critical current densities, than those from the first category, which need.

The preliminary results have shown that a sample of high-T_c ceramic intrinsically consists of superconducting grains connect-

ed by a network of a randon Josephson junctions. Indeed,
a) the "shoulder"-type transformation of transition curve in magnetic field was explained by the hypothesis that at first, under a small magnetic field Josephson junctions were broken and only then granule superconductivity destroyed, b) the microwave induced current steps observed in different kinds of ceramic contacts seemed to prove the importance of inner Josephson junctions (1-7), c) the quasi-periodic magnetic field dependence of the critical current as well as the electrical voltage along a sample in the resistive state (8) was explained by interference phenomena in Josephson junctions contained loops similar to that of Little and Parks investigations (9). Corresponding loops quantization area estimated from this oscillation period about 10^{-1} Oe was proved to be of the order of granule dimensions and many orders larger than the coherence length (8). Thus the existence of macroscopic scale phase coherence in high-T_c ceramics was also demonstrated.

For a practical interference devices like SQUID, the flux quantization area should be defined by loop geometry rather than the sample microstructure. To obtain a practical device made of ceramics a set of problems concerned the Josephson media nature of material should be primarely solved. That is:

1) critical parameters of the weak link, which determines the device operation, should be dumped in comparison with that of the loops volume;

2) the loop quantization area should be sufficiently large to eliminate the influence of its change under external stimulus, for example, this area change might be due to penetration depth temperature dependence, like that observed recently by Gallop et al. (10);

3) the quantization area increase is restricted due to loops inductance L limitations imposed by noise generation arisen from thermal fluctuations. A simple estimation of this intrinsic noise can be obtained under assumption that the source of the fluctuations is the Johnson noise in the normal state resistance of the weak link in the loop. If the obtained under these assumptions

(11) total flux noise:

$$<\Phi_N^2> = k_B TL \geq \Phi_0^2, \text{ where } \Phi_0 = h/2e,$$

the periodicity in the response of the device would be completely wiped out by the noise. Thus a certain loops area optimization is necessary.

These problems were solved in the original paper on flux quantization (12) where the distinct interference patterns in a stable weak link contained ceramic loop with the area of about $10^{-2} cm^2$ were observed at 77K. The period of these patterns of about 10^{-5} Oe was in good agreement with the value of flux quantum calculated for this area. Similar results were observed simultaneously by Zimmerman et al (13) on a loop of about $5 \cdot 10^{-3}$ cm^2 area with a break-junction type weak link. In spite of extremely encouraging value of a spectral density of equivalent flux noise of about 4.5×10^{-4} $\Phi_0/\sqrt{Hz}$ obtained in (13) for this rf-SQUID at 75 K, practical application of such sensor was limited because of break junction instability as well as its degradation during thermal cycling (13). The rf-SQUID studied in (12) have utilized the stable weak link and thus was free from drawbacks mentioned above.

2. Superconducting interferometer operated at liquid nitrogen temperatures.

A "weak link" contained ring made of 123-YBCO ceramic was used as a sensitive element of the rf-interferometer described in (14-15). The hole of 1.5 mm in diameter was made during pellet pressing; "weak link"- a bridge-form narrowing of a rings width to approximately $(10^{-2} cm)^2$ over ~ $10^{-2} cm$ distance was made by mechanical rowing of a sintered specimen. All the measurements were performed using conventional electronics for helium cooled rf-SQUID. Simplified scheme of experimental arrangement is drawn in fig.1. The resonance rf-circuit coil was placed inside the hole in the ring; due to inductive coupling between them it was possible by measuring of circuits characteristics to determine the superconducting transition curve of the "weak link" which coincided with that obtained by four-probe method. "Weak link" transition tempera-

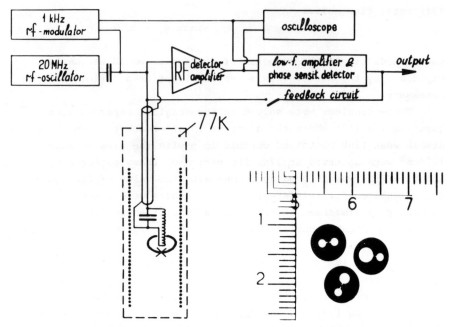

Fig. 1 Schematical drawn of experimental installation; insert: photographs of 123: YBCO-ceramic rf-SQUID sensors

ture was found to be close to that of 123:YBCO material except of the first sensors for which slightly reduced values of about 80 K (as it is seen from fig.2) were obtained. Below this temperature it was possible to switch the "weak link" into a resistive state by means of rf-drive V_{rf} increasing. This event was accompanied by appearence of a periodic magnetic field dependence of the rf-output of device, which was brought out in a customary way by use of low-frequency modulation of magnetic field. Signal-to-noise ratio values as good as 10-30 had been obtained at 77 K (see fig. 3 as example) for approximately 30% sensors from somewhat more that 100 of produced up to nowadays, while the rest displayed the reduced values of about 3-5.

123:YBCO-ceramic made rf-interferometer characteristics obtained experimentally in liquid nitrogen temperature range were similar to those of a customary point contact rf-SQUID operated at liquid helium temperatures. Indeed under high values of signal to noise ratious just like that of Nb-made SQUID:

a) the increase of V_{rf}-drive value, corresponded to the maximum value of them (dots and circuits) with T_c-T and alternation of regions on V_{rf} with and without magnetic field sensitivity as it is shown in fig.2 (it should be mentioned that for the best sensors like that, characterised by fig.3 regions of magnetic field sensitivity absence on V_{rf}-drive proved to be collapsed into dots) and

b) magnetic field dependence of the SQUID detected tank circuit voltage changed its phase during rf-bias increase, as it is shown in fig.3, where first seven periodical patterns displayed at 77 K are reproduced (signal amplitude damping is due to rf-detector nonlinearity) - also confirm this analogy.

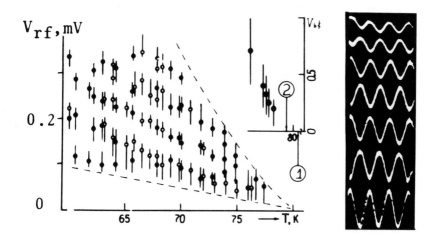

Fig. 2 Temperature dependence of magnetic field sensitive regions on V_{rf}-drive for two sensors

Fig. 3 Magnetic field dependence of detected tank circuit voltage for different values of V_{rf}-drive at 77 K for third SQUID

Unprotected high-T_c ceramic made superconducting rf-interferometer has demonstrated brilliant vitality that is: characteristics of remarkable stability when stored in liquid nitrogen without any noticeable degradation or transformation of them due to numerous thermal cycling with minimal precautions. But still even a brief heating for about 30 min of the sensor to ~ 340 K in a water vapour has erased the superconductivity at 77 K of the "weak link" in agreement with observations of such ceramic decomposition in water made elsewhere (16-17). The influence of this factor as well as an additional noise due to sensor vibration by boiled nitrogen was found to be practically eliminated by hermetic embracement of a sensor into protecting plastical cover. For instance, the one which patterns are reproduced in fig.3 has been withstanding with no detectable signs of degradation more than 10 months test of room temperature storage completed by nearly every day short exposure at 77 K.

The results obtained points out the opportunity of 123-YBCO-ceramic made rf-SQUID production for practical applications, which are determined by brilliant vitality, liquid nitrogen temperature range of operation and characteristics quite similar to those of liquid helium cooled one.

Nevertheless there is a question which should be solved, namely, how such a macroscopic "weak link" with dimensions many orders in magnitude larger than that of not only coherence length, but also of typical granule dimensions can provide characteristics similar to those of point contact device? The simpliest assumption of "break junction" creation in a route of "weak link"rowing seems to be rather doubtful as only a few sensors produced were suspected to have cracks inside "weak link" because of,in spite of enormous large "weak links" dimensions, periodical magnetic field dependences of their rf-output and large signal-to-noise ratious were obtained after first cooling to 77 K but have entailed by a fast degradation of the characteristics in the following thermal cycles (similar to that of(13)),and vice versa - the most of operationable devices were characterised by similar "weak links" dimensions mentioned above and pretty good vitality, i.e. uncovered sensors have withstanded at least 10 thermal cycles without any detactable signs of degradation.

As soon as a rough estimation shows that there are about 10^2 grains in the "weak link" cross section and therefore all the interference phenomena might have been dumped, but they are not; and just as we have rejected "break junction version", so the results obtained can be explained only by assuming that the entire set of Josephson junctions responds as a single one to the external stimulus. Following factors might be responsible for such a behavior:

a) a specific feature of SQUID preparation which results in an identical Josephson junction network formation;

b) a manifestation of mesoscopic effects, which occur due to unsufficient amount of junctions in the "weak link" for complete averaging of their characteristics (see for example Ref.18, Subsection II C.3) or may be simplier,

casual formation of single junction contained cross section which arises due to chaotic junctions distribution in a Josephson media and rather small "weak link" area for this distribution averaging:

c) the appearance in a set of coupled Josephson junctions, upon a variation in V_{rf}-amid the chaos of responses to the external stimulus - of regions (along the V_{rf} scale) of a mutual synchronization of the ensemble, as occurs in nonlinear dynamic systems (19).

Among there possibilities the last two have seemed to be the most likely for us, but a special experimental test was necessary for a proper choice between them.

3. Solitary "weak links" properties

In order to solve the problem, mentioned in the previous section, a single "weak links" properties have been studied (15,20). A-priori it seems evident that depending of realized version, quite different phenomena might be observed. In particular, under microwave radiation patterns contained current steps could be observed: from Josephson equation described equidistant "staircase" for a single junction to the rather complicated one for a multiconnected network of different junctions (21).

"Weak links" were extracted from the stable operationable 123:YBCO -made SQUID-sensors of different quality and were studied with- and without microwave radiation of a frequency range of (25-40)GHz by the four probe method. An uncovered sample was placed at the open end of the 8-mm waveguide attached to the ajustable microwave power source with a set of attenuators and wavelength meters; the metal cover has embraced the cryogenic part of the set up. Experiments were performed in a glassy cryostat with μ-metal schielding in a temperature range from 4.2 to 78 K.

"Weak link" critical current J_c (over 6 samples investigated) in the absence of microwave radiation has achieved $(5-50)10^{-5}$A and $(3-300)10^{-3}$A at 77 and 4.2 K correspondingly and was periodically depended on the external magnetic field. Microwave irradiation of the samples has produced current (Shapiro) steps on the V-I patterns as it is illustrated by fig.4. The positions of these steps depend linearily on microwave frequency in a whole range explored and coincide with the corresponding values calculated from Josephson equation of $V_n = n\hbar\omega/2e$ with a relative precision of $3 \cdot 10^{-4}$. Subharmonical steps at $V = V_n/2$ where sometimes observed as well (left part of fig.6; lower amplitude singularities corresponds to subharmonical structure for odd numbers).

The microwave radiation amplitude A typical dependences of the first four orders Shapiro steps hights obtained at 35.5 GHz are shown in fig.5 by dots connected for clarity by dashed line. These curves differ from the Bessel type because of current source method used and are in a satisfactory agreement with results obtained for a Dayem bridge by analogue computation for this case (22), which are also presented in fig.5 by solid line drawn for $\Omega = \hbar\omega/2eR_nJ_c = 0.02$; and the scale of axes were chosen to fit these results to experiment by J_c to zero linear extrapolation and its 4-th zero. Rather good agreement between experiment and calculations was found to be for the microwave amplitude larger than 0.4; but at lower values of microwave amplitude certain discrepances, probably due to computations approximations, were clearly observed.

This result and the coinsidence within experimental accuracy of microwave induced current steps positions with that calculated from Josephson equation as well as magnetic field dependences period of the SQUID's rf-output with the flux quantum quantity calculated from the loop geometry shows that the "weak link" characteristics are similar to those of a point contact type Josephson junction. Since the low-J_C results were obtained on the "weak links" extracted from the "good" SQUIDs which displayed almost periodical rf-output vs magnetic field patterns, it seems pretty probable that their properties were governed by the microscopic superconducting single junction "narrow neck" situated inside the geometrical "weak link".

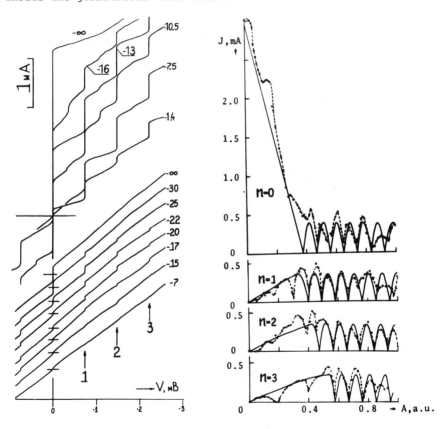

Fig. 4,5 Characteristics of isolated "weak link"

Fig. 4(left): 4.2 & 77 K results (upper & lower parts correspondingly), figures on the curves indicate microwave amplitude attenuation

This conclusion confirms the assumption mentioned above (section 2,b) that due to chaotic junctions distribution inherent to "Josephson medium" and insufficiently large "weak links" dimensions for this chaos averaging - a "narrow neck" junction such that all the superconducting ways connected "weak link" shores are passed through it only, was casually emerged during SQUID preparation. One can estimate also the lower limit of this "narrow neck" cross section by use of a critical current values measured in the absense of the microwave radiation (15,20) which were not less than 3mA at 4.2 K. Since the critical current density for a single crystal is not worse than that for ceramic of the same composition then, by use of a single crystal parameters (23), one can conclude that this cross section can not be less than $10^{-11} cm^2$.

Thereby the properties of 123-YBCO-ceramic made "weak link" are governed (at least for a "good" devices)by a single junction "narrow neck" placed inside with transversal dimensions several orders of magnitude as large as that of coherence length of material (23).

Several interconnected junctions existence in the "narrow neck" seems to be responsible for the high-J_c sample properties obtained. Indeed: a) this sample was extracted from SQUID-sensor which displayed rather complicated far from "triangular" aperiodic patterns slightly reminded that of (24), b)weak link critical current value was almost 100 times as large as that for a "good" device; c) an additional structures included several current steps observed on a V-J curves even in the absense of microwave radiation might be due to manifestation of effects which reminded internally pumped parametric amplifier (25,26) and d) microwave induced current steps assemblage was consisted of almost equidistant steps sets slightly shifted from each other.

At liquid helium temperatures the dozens of Shapiro steps were observed but due to J-V patterns smashing by thermal fluctuations when $hJ_c/2e \leq 5kT$, this amount was reduced to 3-6 at 77 K. Nevertheless, by use of a simple low-frequency modulation technique, we could easily detected microwave induced singularities with numbers as large as 100 and even higher on a frequency of 35.5GHz even at 77 K (see fig. 6 as example). Since these results were obtained by use of a quite primitive installation without any sample - microwave guide ajustment it seems rather probable that such a "weak link" can be used as a sensitive detector, mixer and so on in a frequency range up to $3.5 \cdot 10^{12}$ Hz at liquid nitrogen temperatures. The "weak link" critical current damping under IR-radiation observed at 77 K has besides testified on a probable use of it like a bolometer. Similar results were obtained also elsewhere (27, 28).

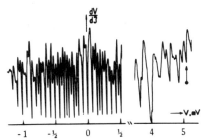

Fig. 6 dV/dJ vs V "weak link's" characteristics at 4.2 (left) and 77 K (right); 35.5 GHz. 70-th step position is indicated by " ↑ "

4. Towards practical applications

The 123-YBCO ceramic made rf-SQUID operated at liquid nitrogen temperatures method of production and reasons for which the macroscopic "weak link" provided the point contact Josephson junction patterns were described in the previous sections. This one is intended for a review of the results of device characteristics investigations from the viewpoint of its possible practical applications to measure external fields and different magnetic measurements like sample susceptibility (29).

Rf-SQUID characteristics were studied for the devices of different geometry as single-hole one, which can immediately be used as an external magnetic field sensor, as well as double-hole one with the holes of different areas separated by the "weak link" (see photographs in fig 1) - intended for the magnetic measurements. Experiments were performed using installation scematically shown in fig.1 in a glassy liquid nitrogen temperature cryostat with a shield made of μ-metal.

The both single-hole rf-SQUID-magnetometers investigated have displayed almost periodical field dependencies of their rf-outputs at 77 K. The periods obtained in the uniform external magnetic field as well as by dc-bias of rf-coil were about 10^{-5} Oe and have corresponded within the error of measurements to the application of one-flux quantum of $\Phi_0 = h/2e$ to the geometrical area of the sensors hole. The best short time scale external field sensitivities of $2 \cdot 10^{-8}$ Oe during 5 min which have been obtained by use of single-hole devices were probably restricted by ambient noise due to inadequate shielding of experimental installation. Thus it was seemed nessesary to obtain a better shielding characteristics of the experimental set up to throw light upon the intrinsic noise level and resolution limit of the device under investigation.

A-priori, double-hole sensor in comparison with single hole one gives an intrinsically lower sensitivity to ambient noise and it is to some extent more mechanically stable. Noise sensitivity reduction apparently depends on the shielding properties of ceramic, the degree of holes asymmetry and pellet width. In order to obtain an estimation of this parameter for the real double-hole devices, their external uniform magnetic field sensitivities were

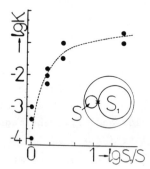

Fig. 7   77 K-external field sensitivity reduction of double-hole sensors in comparison with that of single-hole one

measured. The relative sensitivity reduction K obtained for a set of different sensors as regarded to the sensitivity of the single hole one is drawn in fig.7 as a function of their hole area ratios S_1/S, where S-corresponded value of rf-coil-hole mentioned in Sec. 2. The results obtained have thereby showed that 123 : YBCO-ceramic made double-hole rf-SQUID possessed of lower ambient noise characteristics but still can be used in magnetometry instrumentation.

The sensitivity of $5 \cdot 10^{-4}$ Φ_o to magnetic field produced by dc-bias of rf-coil was obtained for double-hole SQUIDs with S_1 $0.12 cm^2$. It seems thereby evident that the resolution limit of the magnetometer described above is not worse than $5 \cdot 10^{-9}$ Oe. The results obtained are in good agreement with those described previously (14) on the superconducting interferometer with areas ratio of about 1,8. Similar results were obtained recently by Birmingham group (30) on the sensor resembled those described here.

In spite of relatively low way-out of a "good" devices (about 30%) due to casual formation of "narrow neck" and reduced value of their ultimate sensitivity in comparison with that of liquid helium cooled counterparts due to thermal fluctuations induced intrinsic noise enhancement, the 123:YBCO-ceramic rf-SQVIDs produced are seemed to be suitable for a wide area of different applications at liquid nitrogen temperature range.

Asymmetrical double-hole rf-SQUID, as it was mentioned above, can be treated as a flux transformer thus suitable for use to measure (as an example) magnetic susceptibility of a sample placed into a larger hole. The proper shielding of experimental arrangement is nessesary to achieve sufficiently high accuracy level of such a measurements, comparable with that of rf-SQUID utilized. A-priori it seems evident that the most preferable is the use of superconducting shield. Nevertheless the 123:YBCO-ceramics shielding properties were someway investigated (31,32), the used low resolution limit of worse than $2 \cdot 10^{-3}$ Oe was demanded of additional measurements on the enhanced accuracy level.

The shielding properties of a set of screens made of a middling density 123:YBCO-ceramic in a form of a hollow cylinder with a bottom were studied at 77 K using the ceramic rf-SQUID as a sensor (29). The typical screen had the following dimensions: inner diameter of 10 mm, height - 35 mm, walls and bottom width - 1.5 ÷ 2 mm. In order to reduce the influence of magnetic field "hang in" through the open end of the shield, it was closed by a short (about 25 mm in length) screen cover with a small hole for the rf-circuit wires in a bottom. External uniform magnetic field H_e was provided by a long single layer coil; ceramic rf-SQUID-magnetometer placed inside the screen in 5-7 mm above its bottom was measured the "internal" field H_i. The right part of fig. 8 shows a typical external magnetic field dependence of the magnetometer output, proportional to the value of H_i inside the shield. The results obtained in both alternating (at a frequency of 0.1 Hz) and constant field are drawn by dashed and solid lines correspondingly. Magnetic field inside the screen has been found to conserve its value within long time accuracy limit of not worse than 10^{-7} Oe, during external field variation in a range of 1-9 Oe for 4 screens investigated. It should be pointed out that the shielding coefficient reduction on a factor of 10-100 was observed due to magnetic field "hang in" effects manifestation in the case of uncovered shield.

When the external field value corresponded to complete shielding was exceeded (but still much less, than that of H_{c1} of material) it had penetrated inside the screen probably due to intergrain vortex motion. This phenomenon was characterised by a rather complicated time dependences as it is shown in the left part of Fig.8. The established value of H_i was achieved after a proper time depended on external field; for example, it has been reached 300 min in a field of 4 Oe for the shield characterised by the patterns drawn in the insert to fig.8. Further internal field variation was undetectable in the long time resolution limit of our installation. Such steady state field H_i has been found to conserve its value during external field reduction in the range of 1-9 Oe (see the right part of Fig.8).

The results discussed show that even such a middling quality shield can be used in a route of low field magnetisation properties investigations at liquid nitrogen temperatures. Here we

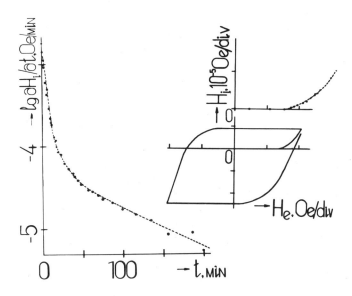

Fig.8 123:YBCO-ceramic made shield characteristics studied at 77K by use of high-T_c rf-SQUID

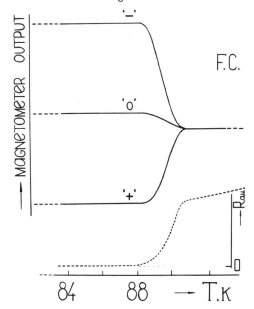

Fig.9 Transition curves of 123:YBCO-ceramic sample obtained by measure of dc-resistivity as well as magnetic moment

present the preliminary results of this work from the viewpoint of ceramic rf-SQUID application mainly.

The temperature dependence of 123:YBCO-ceramic magnetization was measured at the quantum flux level by double-hole rf-SQUID. A sample in the form of cylinder with dimensions: 1 mm in diameter and 4 mm - height, with additional electrical leads for DC-measurements was placed inside 2.7 mm in diameter hole of the SQUID. The sample temperature was measured by the thermocouple and could be ajusted by heating of a copper pivot to which it was attached by In-solder. Superconducting shield with a vacuum experimental chamber inside was maintained at liquid nitrogen temperature. Thermal isolation between rf-SQUID sensor and the sample was sufficient to allow the specimen to be heated up to 110 K before SQUID temperature change became noticeable. Magnetic field was produced by a single layer copper coil, the measurements were performed in a trapped flux regime.

Typical ceramic transition curves obtained by four-terminal potentiometrical method and by magnetic moment measurements are shown in fig. 9 by dashed and solid lines correspondingly. Each solid line corresponds to a set of thermal cycles during which no detectable sign of magnetic moment transition curves irreproductibility or hysteresys have been observed within the experimental accuracy. Weak variation of specimens magnetic moment in a superconducting state probably due to penetration depth temperature dependence as well as flux jumps during transition were out of Fig. 9 scale. By use of additional calibration coil placed into the samples hole of SQUID, Meissner effect value of about 40% for the sample explored was estimated from the data obtained. It should be mentioned that this value might be enhanced because of possible field trapping by the specimen, which was undetectable in the flux trapped regime utilized. The residual Earth field axial component value of 3 mOe was estimated in a self-consistent manner from the data obtained, with addition of external field of 16 mOe with opposite directions which are marked by +/- in fig. 9 as well as that corresponded to the absence of coil-produced field, marked by zero. During these tests the values of data obtained were at least four orders in magnitude as large as the resolution limit

of the experimental set up, it thereby allows to estimate the scope
of potential utilization for a liquid nitrogen temperature range
applications of double-hole rf-SQUID even in the absence of wire
made of high-T_c material.

5. Conclusions

123:YBCO-ceramic "weak link" with typical dimensions of
$(10^{-2} cm)^2$ produced have displayed characteristics similar to that
of a point contact Josephson junction in an interference contour
as well as under microwave radiation. These effects were governed
by microscopic single junction contained "narrow neck" formation
inside the "weak link" during its preparation route.

The opportunity of practical realisation of mixers, frequency multipliers and converters as well as detectors made of high-T_c ceramic and operated in a range up to $3.5 \cdot 10^{12}$ Hz at liquid
nitrogen temperatures - was confirmed by observation at 77 K of
the superior order Shapiro steps with numbers as large as 100 and
even higher obtained during "weak link" investigation for microwave frequency of 35.5 GHz.

Single- and double-hole rf-SQUIDs, based on utilisation of a
"weak link" displayed at liquid nitrogen temperatures almost
periodical magnetic field dependences of device rf-output patterns
quite similar to that of a point contact liquid helium cooled
counterpart, have been produced and investigated. The periodicity
of the modulation patterns obtained has agreed with the flux
quantum value expected from geometry. External magnetic field
sensitivity of $2 \cdot 10^{-8}$ Oe obtained with single-hole devices was
restricted by ambient noise, while for intrinsically better shielded double-hole rf-SQUID the equivalent flux noise of $5 \cdot 10^{-4} \Phi_o$ has
been obtained. It seems quite reasonable that a certain improvement of parameters obtained might be achieved by perfection of
device production technology as well as electronics and noise
reduction. Brilliant stability of characteristics and high signal-to-noise ratio level as well as inaffection of thermal cycling
and long time storage on them - emphasizes the attractiveness of
high-T_c-ceramic rf-SQUIDs for use in a different applications in

a liquid nitrogen temperature range, as is illustrated by a set of experiments produced by use of these devices as a sensors.

Concerning dc-SQUIDs, which were out of this paper framework, it should be mentioned that in the case of device made of high-T_c-film in spite of the extremely low flux noise figures up to $10^{-5} \phi_0/\sqrt{Hz}$ obtained on a quantization area of about $10^{-4} cm^2$ (33), an enormous amount of technological problems should be solved (27,30,34) to obtain sensors suitable for practical applications.

References

(1) J.S.Tsai, Y.Kubo and J.Tabuchi. Phys. Rev. Lett. 58,(1987)1979.
(2) A.V.Varlashkin, A.L.Vasil'ev, A.I.Golovashkin, O.M.Ivanenko, N.A.Kiselev, L.S.Kuz'min, K.K.Licharev, K.V.Mizen, G.V.Romanchikova and E.S.Soldatov, JETP Lett. 46, (1987) S52.
(3) B.I.Verkin, S.I.Bondarenko, V.M.Dmitriev, A.V.Lukashenko, V.P.Seminozhenko and A.A.Shablo,Sov.J.Low Temp. Phys. 13 (1987) 995.
(4) J.Moreland, L.F.Goodrich, J.W.Ekin, T.E.Capobianco, A.F.Clark, A.I.Braginski and A.J.Panson, Appl. Phys.Lett. 51 (1987) 540.
(5) S.Kito, H.Tanabe, T.Takahashi, M.Tonouchi, Y.Fujiwara and T.Kobajashi, Jap. J.Appl.Phys. 26, (1987) 1353.
(6) V.Higashino, T.Takahashi, T.Kawai and S.Naito. Jap. J.Appl. Phys. 26(1987) 1211.
(7) D.Esteve, J.M.Martinis, C.Urbina, M.H.Devoret, G.Collin, P.Monod, M.Ribault and R.Revcolevshi, Europhys.Lett.3 (1987)1237.
(8) N.V.Zavaritsky, V.N.Zavaritsky, S.V.Petrov and A.A.Jurgens JETP Lett. 46 (1987) S18.
(9) W.A.Little and R.D.Parks, Phys. Rev.Lett. 9 (1962) 9.
(10) J.C.Gallop, C.D.Landham, W.J.Radcliffe and W.B.Roys, Physica C153-155 (1988) 1403.
(11) J.E.Zimmerman, Cryogenics 12 (1972) 19.
(12) N.V.Zavaritsky, V.N.Zavaritsky and S.V.Petrov, Novel Superconductivity (S.Wolf and V.Kresin, Plenum, N.Y.1987)871.
(13) J.E.Zimmerman, J.A.Beall, M.W.Cromar and R.H.Ono, Appl.Phys. Lett. 51 (1987) 617.
(14) N.V.Zavaritsky, V.N.Zavaritsky and S.V.Petrov, JETP Lett.46

(1987) 592.

(15) N.V.Zavaritsky and V.N.Zavaritsky, Physica C153-155 (1988) 1405.

(16) I.Nakado, S.Sato, Y.Oda and I.Kohara, Jap. J.Appl. Phys. 26 (1987) 697

(17) M.F.Fan, R.L.Barns, H.M.O'Bryan Jr, P.K.Gallagher, R.S.Sherwood and S.Jin, Appl.Phys.Lett. 51 (1987) 532.

(18) A.G.Aronov and Ju.V.Sharvin, Rev.Mod.Phys. 59 (1987).

(19) M.I.Rabinovich, Sov.Phys.Usp. 21, (1978) 443.

(20) N.V.Zavaritsky and V.N.Zavaritsky, Sov.JETP Lett. 47 (1988) 334.

(21) T.D.Clark, Phys. Rev. B8 (1973) 137.

(22) P.Russer, J.Appl.Phys. 43 (1972) 2008.

(23) T.R.Dinger, T.K.Worthington, W.J.Callagher and R.L.Sandstrom, Phys. Rev. Lett. 58 (1987) 2687.

(24) V.M.Zakosarenko, E.V.Il'ichev, T.N.Nikiforova and V.A.Tulin, Sov.J.Tech.Phys.Lett. 13 (1987) 1389.

(25) S.Wahlsten, S.Rudner and T.Claeson, J.Appl. Phys. 49 (1978) 4248.

(26) A.N.Vystavkin, V.N.Gubankov, G.F.Leshchenko, K.K.Licharev and V.V.Migulin, Radio Engin.Electron.Phys 15 (1970) 2121.

(27) U.Kawabe, Physica C153-155 (1988) 1586.

(28) Y.Enomoto, T.Murakami and M.Suzuki, Physica C153-155(1988)1592.

(29) N.V.Zavaritsky and V.N.Zavaritsky. Sov.J.Tech.Phys.Lett. 14 (1988) in press.

(30) C.E.Gough, Physica C153-155 (1988) 1569.

(31) V.Vvedenskii, I.Graboy and A.Kaul', High T_C-Superconductivity Research 1 (V.I.Ozhogin, Kurchatov Inst.of Atomic Energy Moscow,1987), 78.

(32) J.C.Macfarlane, R.Driver, R.B.Roberts, E.C.Horrigan and C.Andrikidis, Physica C153-155 (1988) 1423.

(33) D.Robbes, Y.Monfort, M.Lam Chok Sing, D.Bloyet, J.Provost, B.Raveau, M.Doisy and R.Stephan, Nature 331 (1988) 51.

(34) R.H.Koch, C.P.Umbach, M.M.Oprisko, J.D.Mannhart, B.Bumble, C.J.Clark, W.J.Gallagher, A.Gupta, A.Kleinsasser, R.B.Laibowitz, R.B.Sandstrom and M.R.Scheuermann, Physica C153-155 (1988) 1685.

HIGH-T_c SUPERCONDUCTING THIN-FILM WEAK LINKS: JOSEPHSON EFFECT AND MACROSCOPIC QUANTUM INTERFERENCE AT 4 K AND 77 K

A.I. Golovashkin,[+] A.L.Gudkov,[++] S.I. Krasnosvobodsev,[+]
L.S.Kuzmin,[++] K.K.Likharev,[++] Yu.V.Maslennikov,[++]
Yu.A.Pashkin,[+] E.V.Pechen,[+] and O.V.Snigirev[++]
[+] Lebedev Institute of Physics, U.S.S.R. Academy of Science, Moscow 117924, U.S.S.R.
[++] Department of Physics Moscow State University, Moscow 119899 GSP, U.S.S.R.

Abstract

We have studied properties of single and double superconducting weak links ("Dayem" microbridges) fabricated by direct photolitography from thin films of 1-2-3 copper oxides, deposited on $SrTiO_3$ and $LiNbO_3$ substrates by laser sputtering of ceramics targets. Values of critical current I_c and normal resistance R_n of the weak links could be altered by plasma etching of the structures. $I_c R_n$ products up to 100 μV were achieved and well pronounced Shapiro steps under 8 GHz microwave radiation were observed at 77 K. Macroscopic quantum interference was registered in double microbridge structures at the same temperature.

Introduction

In a short time after discovery of the high-T_c superconductivity of the copper oxides, the Josephson effect has been proved to take place in weak links of the new superconductors.[1-3] Nevertheless, numerous attempts to fabricate reproducible Josephson junctions with high values of the basic parameters, notably the characteristic frequency

$$\omega_c = \frac{2\pi}{\Phi_0} I_c R_n \qquad (1)$$

have not been very successful so far.

Fortunately, the Josephson effect has been demonstrated to take place in natural junctions between crystallites ("grains") of both bulk[3-5] and thin-film[6-11] ceramics of the high-T_c superconductors. Although parameters of such junctions could hardly be reproducible, they can be useful for simple Josephson-effect-based devices like SQUIDs at the first stage of development of the high-T_c superconductor electronics.[12]

The purpose of this work has been to study experimentally basic properties of the natural intergrain Josephson junctions in both granular and well-oriented thin films of 1-2-3 phase superconductors.

Thin Films

Thin films of $YBa_2Cu_3O_{7-\delta}$ and $HoBa_2Cu_3O_{7-\delta}$ were deposited by laser sputtering of bulk ceramic targets of these materials on heated substrates of 100-oriented $LiNbO_3$ and $SrTiO_3$. A solid state laser produced 10 ns pulses with the repetition frequency 50 Hz, so that deposition of 0.5 - 1.0 μm thick films took about one minute.[13,14]

X-ray diffraction studies have shown that the films deposited on the $SrTiO_3$ substrates were totally oriented, with c axis normal to the surface. Normal resistivity of the films (at 100 K) was below 100 μΩ-cm; they exhibited a sharp resistance transition to the superconducting state at T_c =91 ± 1 K. Critical currents of these oriented films were as high as $\sim 10^6$ A/cm^2 at 77 K; such currents allowed us to demonstrate magnetic levitation of the substrate with the deposited film in liquid nitrogen.[14]

On the contrary, the films deposited on $LiNbO_3$ substrates under similar conditions were found to consist of randomly oriented grains of ~ 1 μm size. Their resistivity was larger and critical currents were lower ($10^4 - 10^5$ A/cm^2) that those of the well-oriented films.

Weak Links

In order to single out an intergrain Josephson junction we have patterned our films to form uniform-thickness ("Dayem") microbridges with planar dimensions 10×10, 10×2 and 2×2 μm^2. In contrast with earlier works where more sophisticated methods of patterning of high-T_c films were employed,[6-10] we used quite standard direct photolithography. After deposition and exposure, the 1-μm-thick photoresist AZ-1350 was dried at 70°C for one hour and then developed in 1% water solution of KOH. Then the superconducting film was etched in 1% water solution of HNO_3. A thinner film of unknown composition remained on the etched areas (Fig.1); its sheet conductance was, however, much lower than that of the unexposed parts of the superconducting film. Properties of the latter

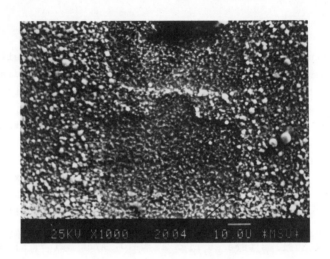

Figure 1. Scanning electron micrograph of a weak link ("Dayem microbridge") fabricated from granular $Y_1Ba_2Cu_3O_{7-x}$ thin film deposited on the $LiNbO_3$ substrate

film were virtually unaffected by the lithography; in particular, reduction of its T_c was never larger than 2-3 K.

Critical current I_c of even the smallest microbridges patterned from well-oriented films were usually so high (few mA) that effects of self-heating were observable. In order to reduce I_c, we processed the structures in rf plasma discharge of O_2 with gas pressure close to 5×10^{-2} Torr. The substrate temperature was about 50 °C, dc voltage developed at the substrate holder was 100--150 V. The processing resulted in gradual decrease of I_c (Fig.2) and increase of R_n of the microbridges in such way that the I_cR_n products remained approximatelly intact, apparently due to decrease of thickness of the superconducting film. After completion of the fabrication process, the sample was coated by the same photo-resist in order to prevent its exposure to the water vapor.

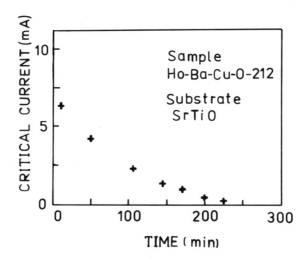

Figure 2. Critical current of the microbridge at 77 K as a function of period of its processing in the oxigen plasma.

DC I-V Curves

Figure 3 shows typical dc I-V curves of the weak link formed using granular film (on $LiNbO_3$ substrate), recorded at the helium and nitrogen temperatures. One can see that $I_c R_n$ products of the junctions can be as high as $\sim$ 500 µV at 4 K and $\sim$ 100 µV at 77 K. The both figures are close to the highest values reported for high-T_c weak links.[9,10,17] Shape of the I-V curves (no hysteresis, large excess current $I_{ex} \simeq I_c$) is typical for S-N-S Josephson junctions.[15] A 8-GHz microwave irradiation of the samples induces Shapiro steps separated by usual intervals $\Delta V = (2e/\hbar) \simeq 16$ µV. Widths of the steps as functions of microwave power do oscillate even at helium temperatures, giving an evidence of a nearly sinusoidal $I_s(\Phi)$ relationship.[15]

At nitrogen temperatures, the I-V curves are noticeably "washed out" by fluctuations. This effect is somewhat larger than that following from the standard theory of the thermal fluctuations [14,15] because of reasons unknown so far.

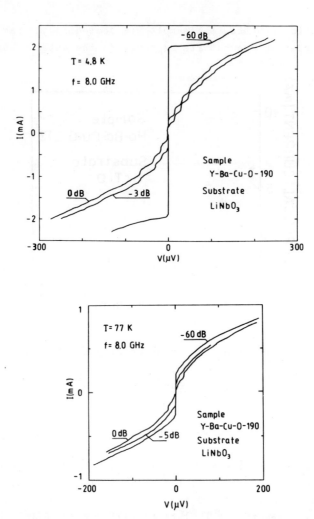

Figure 3. DC current-voltage characteristics of a weak link (10×10 µm^2) formed from granular film on the LiNbO$_3$ substrate for various relative levels of applied microwave power: (a) at helium and (b) nitrogen temperatures

Somewhat unexpectedly, very similar results have been obtained using well-oriented (apparently quasi-monocrystalline) films deposited on $SrTiO_3$ substrates -see Fig.4. This fact implies that critical currents of even such "monocrystalline" films are determined by the Josephson junctions formed between their crystallites (blocks) rather than by their internal properties (flux pinning, etc.). This conclusion is in agreement with that of the recent work.[10]

Macroscopic Quantum Interference

Except single microbridges, we have fabricated also superconducting thin-film loops of diameter ∼ 50μm with two similar microbridges. These superconducting quantum interferometers exhibited (both at 4 K and 77 K) quasiperiodic modulation of their critical currents by magnetic field normal to the plane of the film. Modulation depth ΔI_c was close to the theoretical value Φ_0/L, where $L \simeq 0.1$ nH is the loop inductance. Magnetic field period ΔH of the interference pattern was of order of $(S/\Phi_0)^{-1}$, where S is the loop area; more exact comparison taking into account diamagnetic effect of the thin film is still to be carried out. These observations imply that the reason of the modulation is really the macroscopic quantum interference along the main contour of the loop rather than along other random contours inside the superconducting film.

Both ΔI_c and ΔH did not change as a result of additional plasma etching of the microbridges. However, after a certain stage of the etching process the modulation disappeared. We are interpreting this fact as complete destruction of superconductivity in one of the weak links. A way to ensure simultaneous reduction of I_c of the both weak links is not found yet; at this stage we were able to increase R_n of the interferometer to 0.2 Ohm before destruction of the interference; at this value the dc voltage modulation depth $\Delta V \simeq \Delta I_c R_n$ remains too small (∼ 4 μV) to use the interferometers in practical dc SQUIDs.

Conclusion

Our observations confirm that S-N-S type Josephson junctions are naturally formed between crystallites of both granular and

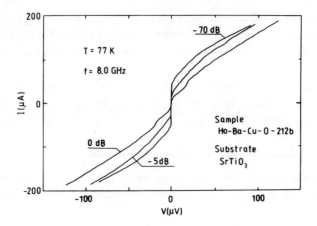

Figure 4. DC current-voltage characteristic of a weak link (2×2 μm^2) formed from well-oriented film on SrTiO$_3$ substrate for various relative levels of applied microwave power at nitrogen temperature.

well-oriented thin films of high-T_c superconductors. $I_c R_n$ products of the junctions can be at least as high as $\sim$ 500 μV at helium temperatures and $\sim$ 100 μV at nitrogen temperatures, while their $I_s(\Phi)$ relationships are single-valued and close to sinusoidal ones. With routine photolithography, one can single out such junctions and use them in simple devices based on the Josephson effect, operating at the nitrogen temperatures.

Acknowledgment

Helpful assistance by E.S.Soldatov and A.N.Tavkhelidze is gratefully acknowledged.

References

1. T.Yamashita, A. Kawakami, T. Nishihara, Y. Hirotsu and M. Takata, Jpn.J.Appl.Phys.,vol.26, p. L635, May 1987.
2. J. S. Tsai, Y. Kubo, and J. Tabuchi, Phys.Rev.Lett., vol. 58 pp. 1979-81, May 1987.
3. A. V. Varlashkin, A. L. Vasiliev, A.I. Golovashkin, O. M. Ivanenko, N.A. Kiselev, L. S. Kuzmin, K. K. Likharev, K. V. Mit-

sen, G. V. Romanchikova, and E. V. Soldatov, Report at the Superconductivity Conference of Low Temperature Physics Counsil of the USSR Academy of Science, Moscow, May 25, 1987 (published in Supplement to Pis'ma v Zh.Teor.Exp.Fiz, JETP Lett. vol. 46, pp. 59-62).

4. D. Esteve, J. M. Martinis, C. Urbina, M. N. Devoret, G. Collin, P. Monod, M. Ribault, and A. Revcoleschi, Europhys.Lett., vol. 3 pp. 1237-42, June 1987.

5. Y. Higashino, T. Takahashi, T. Kawai, and S. Naito, Jpn.J.Appl. Phys., vol. 26, pp. L1211-13, July 1987.

6. R. H. Koch, C. P. Umbach, G. J. Clark, P. Chaudhari, and R. B. Laibowitz, Appl.Phys.Lett., vol. 51, pp. 200-02, July 1987.

7. H. Nakane, Y. Tarutani, T. Nishino, H. Yamada, and U. Kawabe, Jpn.J.Appl.Phys., vol. 26, pp. L1925-26, November 1987.

8. H. Tanabe, S. Kita, Y. Yoshihiro, M. Tonoushi, and T. Kobayashi, Jpn.J.Appl.Phys., vol.26, pp L1961-63, December 1987.

9. B. Häuser, M. Diegel, and H. Rogalla, Appl.Phys.Lett., vol.52, pp. 844-46, March 1988.

10. P. Chaudhari, J. Manhart, D. Dimos, C. C. Tsuei, J. Chi, M. M. Oprysko, and M. Scheurmann, Phys.Rev.Lett., vol.60, pp.1653-56, April 1988.

11. A. I. Golovashkin, A. L. Gudkov, S. I. Krasnosvobodtsev, L. S. Kuzmin, K. K. Likharev, Yu. V. Maslennikov, Yu. A. Pashkin, E. V. Pechen, and O. V. Snigirev, Pis'ma v Zh.Tekhn.Fiz. [Sov.Techn.Phys.Lett.], vol. 14, pp.1256-1259, July 1988.

12. K.K. Likharev, V.K. Semenov, and A.B. Zorin, in: Modern Superconductor Devices, ed. by S.T. Ruggiero and D.A. Rudman, Boston: Academic Press, to be published; see also Report ED-19, presented at 1988 Applied Superconductivity Conference, August 1988.

13. A.I. Golovashkin, E.V. Ekimov, S.I. Krasnosvobodtsev, E.V. Pechen, and V.V. Rodin, Report at conference cited in Ref. 3, pp. 200-03.

14. A.I. Golovashkin, E.V. Ekimov, S.I. Krasnosvobotsev, and E.V. Pechen, Pis′ma v Zh.Exp.Teor.Fiz.[JETP Lett.] vol.47, pp.157-59, February 1988.

15. K.K. Likharev, Rev.Mod.Phys., vol. 59, pp.101-160, February 1979.

16. K.K. Likharev, "Dynamics of Josephson Junctions and Circuits". New York: Gordon and Breach, 1986, ch.4.

17. T. Yamashita, A. Kawakami, s. Noge, W. Xu, M. Takata, T. Komatsu, and K. Matusita, Report ED-18, presented at 1988 Applied Superconductivity Conference, August 1988.